Fundamentals of Finslerian Diffusion with Applications

Fundamental Theories of Physics

An International Book Series on The Fundamental Theories of Physics: Their Clarification, Development and Application

Volume 101

Fundamentals of Finslerian Diffusion with Applications

by

P. L. Antonelli
Department of Mathematical Sciences,
University of Alberta,
Edmonton, Alberta, Canada

and

T. J. Zastawniak
Department of Mathematics,
University of Hull,
Kingston upon Hull, U.K.

KLUWER ACADEMIC PUBLISHERS
DORDRECHT / BOSTON / LONDON

Fundamentals of Finslerian Diffusion with Applications

by

P. L. Antonelli

Department of Mathematical Sciences,
University of Alberta,
Edmonton, Alberta, Canada

and

T. J. Zastawniak

Department of Mathematics,
University of Hull,
Kingston upon Hull, U.K.

KLUWER ACADEMIC PUBLISHERS

DORDRECHT / BOSTON / LONDON

A C.I.P. Catalogue record for this book is available from the Library of Congress.

ISBN 0-7923-5511-3

Published by Kluwer Academic Publishers,
P.O. Box 17, 3300 AA Dordrecht, The Netherlands.

Sold and distributed in North, Central and South America
by Kluwer Academic Publishers,
101 Philip Drive, Norwell, MA 02061, U.S.A.

In all other countries, sold and distributed
by Kluwer Academic Publishers,
P.O. Box 322, 3300 AH Dordrecht, The Netherlands.

Printed on acid-free paper

Contents

Introduction

The erratic motion of pollen grains and other tiny particles suspended in liquid is known as *Brownian motion*, after its discoverer, Robert Brown, a botanist who worked in 1828, in London. He turned over the problem of why this motion occurred to physicists who were investigating kinetic theory and thermodynamics, at a time when the existence of molecules had yet to be established. In 1900, Henri Poincaré lectured on this topic to the 1900 International Congress of Physicists, in Paris [Wic95]. At this time, Louis Bachelier, a thesis student of Poincaré, made a monumental breakthrough with his Theory of Stock Market Fluctuations, which is still studied today, [Coo64].

Norbert Wiener (1923), who was first to formulate a rigorous concept of the Brownian path, is most often cited by mathematicians as the father of the subject, while physicists will cite A. Einstein (1905) and M. Smoluchowski. Both considered Markov diffusions and realized that Brownian behaviour could be formulated in terms of parabolic 2nd order linear p.d.e.'s. Furthermore, from this perspective, the covariance of changes in position could be allowed to depend on the position itself, according to the invariant form of the diffusion introduced by Kolmogorov in 1937, [Kol37]. Thus, any time-homogeneous Markov diffusion could be written in terms of the *Laplacian*, intrinsically given by the symbol (covariance) of the p.d.e., plus a drift vector. The theory was further advanced in 1949, when K. Yosida was able to study Brownian motion on a 3-sphere using semi-group methods, [Yos49]. This linear operator approach was fundamental to the Russian school of E.B. Dynkin and led to many interesting theorems connecting geometry of the domain to properties of diffusions, [Dyn65].

Soon after in the 1940's and 1950's the subject of *Stochastic Differential Equations* was developed by K. Itô and his school and was used extensively in physics, engineering and finance, according to the method of "adding white noise," or simply, the *noise ansatz.* In this period, Itô and McKean's book on the Brownian path appeared, setting a high standard in the subject, [IM64].

Then came McKean's book on stochastic (Itô) integrals [McK69], the books of N. Ikeda and S. Watenabe [IW89], and finally that of K.D. Elworthy [Elw82]. They allowed for a systematic study of *Stochastic Riemannian Geometry.* Such subjects as Feynman-Kac solutions to PDE's (Chapter 4) and the role of Riemann scalar curvature, $\mathcal{R}$, in the Onsager-Machlup Theory (Chapter 5) could then be used in applications in science and engineering, on problems such as filtering and large deviations [AES87], [Var84]. Through the 60's, 70's and 80's Stochastic Riemannian Geometry grew into a vast modern edifice because of the efforts of the Itô and Dynkin schools. For up to date reviews see the papers of Mark Pinsky [Pin78], [Pin92].

By the late 1980's, Stochastic Riemannian Geometry had been applied to various subjects, for example, nonequilibrium thermodynamics, [Gra77b], [Gra77a], and Stochastic Nelson Mechanics, [Nel85]. Yet the work of some of the contributors of this book on coral reef ecology and chemical warfare in plants and marine invertebrates [AS92], [AKR+88], [AL90], [Ant90] marked a return to the biological world from which Brownian motion had originally come. That reversion originated from Peter Antonelli's reading a paper of N. Nagasawa (1980) on spatial patterns in highly *social* and mobile monkey troops, [Nag80], which prompted an interest in Nelson's stochastic (conservative diffusion) approach to Quantum Mechanics and its application to invertebrate developmental models. Furthermore, it was through the study of *social interactions* in ecology that the importance of Finsler geometry was first realized, [ABL91], [ABL92], [ABKS93]. The breakthrough can be phrased in terms of Volterra-Hamilton systems, [AIM93], [AB96].

Letting (X^i, N^i) denote the natural phase space coordinates in a local chart of the tangent bundle TM, consider the 2nd order system

$$\sum : \begin{cases} \dfrac{dX^i}{dt} = k_{(i)}N^i \quad \text{(not summed)} \\ \dfrac{dN^i}{dt} = -\Gamma^i_{jk}N^jN^k + r^i_jN^j + e^i, \end{cases}$$

where all coefficients (possibly) depend on X^i, N^i, t; the n^3 functions Γ^i_{jk} are homogeneous of degree zero in the N^i; and with smooth initial conditions X^i_0, N^i_0, t_0. For almost twenty years this system has played a major role in mathematical theories of ecology, evolution and development in colonial invertebrates such as corals, starfish, bryozoans and other marine fauna, [ABKS93], [AKR+88], [AL90]. The coordinates X^i are Volterra's production variables, whose constant per capita rate of increase is k_i, while the second part of the system is a description of how different species or subpopulations of a colonial organism (i.e. castes) $N^i \geq 0$, *grow* (r^i_j), *interact*

(Γ^i_{jk}) and *react* (e^i) to external influences. Generally, this scheme must entail competition, symbiosis or parasitism, while predator effects usually require additional equations coupled to $\sum$. The condition that Γ^i_{jk} are functions of *ratios* of N^i, mentioned above, signals the presence of so-called *social interactions*, which are higher-order, *density-dependent effects.* Whereas classical theory would have the Γ^i_{jk} merely constants, the theory of density dependent social interactions, initiated by the great ecologist G.E. Hutchinson in 1946, [Hut78], [Hut47], found experimental verification in subsequent work of Wilbur, Hairston and others, [Wil71b], [Wil71a], [HAC+69]. In 1991, work with marine biologist Roger Bradbury indicated that Hutchinson's theory must be recast using zero degree homogeneous interactions (i.e. the Γ^i_{jk}), in order to be consistent with this data. Unfortunately, the mathematical approach that Hutchinson had used in 1946 was intractable and his theory lay fallow for more than 40 years. Thus, 1991 marks the birth of a mathematically accessible theory of Hutchinsonian social interactions, and also the realization that *Finsler geometry describes cost-effective growth and physiology in socially interacting colonial organisms* like siphonophores, ants and other social insects, as well as many species of (Acropora) corals. It also can be applied to the myxomatosis disease epidemic model (Chapter 4).

In order to model environmental noise in the Volterra-Hamilton system $\sum$, it is not enough to merely add white noise to the second part, or population equations. The fact can be simply illustrated with a coral colony. This grows mainly due to the sun's energy which induces photosynthesis in each coral polyp's endosymbiotic algae, allowing the whole colony to produce a calcium carbonate exoskeleton. The coral colony is very sensitive to sunlight, so that random fluctuations in cloud cover have a striking effect on a colony's exoskeleton production (the X's), causing erratic variations. The point of this example is that *both* portions of $\sum$ must be augmented with noise. The question is how to do this. Somehow, the noise term added to the first part, or production equations, must reflect the interactions through which the fluctuations are propagated, for these interactions express the physiology which ultimately produces the calcium carbonate. Our answer to this is general enough to apply to all uses of the Volterra-Hamilton model, and not only the coral example.

The *noise ansatz* we use for the population equations, for the case of *no social interactions* (i.e. Γ^i_{jk} do *not* depend on ratios of population sizes N^i/N^j), is just the usual addition of white noise of constant variance. On the other hand, *a point in production space* (X's) *will be displaced by a distance proportional to the magnitude of the perturbation.* But where does the concept "distance", used here, derive from? The answer is that growth and physiology are generally cost-effective (or nearly so) so that it is natural to use an a priori given cost-functional whose associated Euler-Lagrange equations will provide the coefficients Γ^i_{jk} (but not generally the other terms) in the $\sum$ equations.

In the case of the Finsler cost-functional needed to obtain *cost-effective social* interactions (the Γ^i_{jk} depend explicitly on ratios of N^i), the Finsler distance function is used in our noise ansatz in place of a Riemannian one called for in the absence of social interactions. Also, the *noise* added to the population equations of $\sum$ is *Minkowskian*, rather than Euclidean or white. Thus, our *noise ansatz* explains why all constructions take place on the tangent bundle TM or one of its associated bundles. This is the natural setting for Finsler geometry, as developed by R. Miron and his school, [MA94], [AM96]. It is here that we develop *Finslerian Diffusion Theory* (Chapter 3) to be applied in Chapter 4 to Volterra-Hamilton systems with noise for both the nonsocial (Riemannian) and social (Finslerian) types. There is in that chapter a detailed discussion of E. Nelson's conservative diffusion theory, and *filtering problems for Riemannian Volterra-Hamilton systems* are discussed in detail, while the full power of *Finslerian Diffusion Theory* on TM is brought to bear on the *problem of myxomatosis*, the European Wild Rabbit disease.

In Chapter 5 we prove an *Onsager-Machlup formula* for a class of Finsler spaces which arise in applications in the theory of evolution in biology, [AZ97]. These are the so-called Berwald spaces with positive definite Finsler metric tensor. Such formulas in the Riemannian case have involved the *Riemannian scalar curvature*: for example, the $\frac{R}{12}$ term in the asymptotic sojourn time formula for a Riemannian diffusion, [TW81], or in the nonequilibrium statistical thermodynamics [Gra77b], [Gra77a]. Also, *Riemannian scalar curvature* often enters *non-linear filtering theory*, through the Zakai equations. In fact, this is shown to be always true for 2-dimensional signals in Chapter 4.

Our *Onsager-Machlup formula* for positive definite Berwald spaces (the best understood Finslerian spaces) *applies to h-diffusions*, only. We first formulate a stochastic version of kinetic energy, called *quadratic dispersion*, which in normal coordinates has the form

$$\frac{1}{2}g_{ij}(x_s, y_s)E_s\{x^i_{s+h}x^j_{s+h}\}$$

where $E_s\{\cdot\}$ is the conditional expectation given (x_s, y_s), where (x, y) is a Finslerian h-diffusion. Expanding in powers of h, one obtains $\frac{R}{12}$ as the coefficient of the quadratic term. R is the Gauss-Berwald scalar curvature, which generalizes the familiar Riemannian scalar curvature invariant to Finsler spaces. This result has not been generalized to more complicated Finsler spaces, largely because our proof, like all known proofs of Onsager-Machlup formulas, depends on the existence of normal coordinates and these exist only in Berwald spaces [Dou28], [AIM93]. Yet, this class of Finsler spaces is important in applications, for example, in coral reef ecology problems such as the crown-of-thorns starfish devastation of the Great Barrier Reef, [AS92], [AKR+88], [AL90], [Ant90]. An example of how the *Onsager-Machlop term enters a Finslerian filtering problem* is given in Section 5.4. In Section 5.5 its role in *entropy production* is examined in detail.

In Chapter 6 we answer the question of how the Finslerian hv-Brownian motion is related to the Laplace-Beltrami operator on TM, provided with the Riemannian metric defined by the Sasaki lift (i.e. diagonal lift) of the Finslerian metric tensor, $g_{ij}(x, y)$. For the Finslerian Laplacian expressed in Riemannian terms, one should always add to Laplace-Beltrami operators drift fields, coming from two torsion tensors intrinsic to any Finsler geometry, not both of which can be vanishing. The Finslerian Laplacian is, however, *intrinsically defined* by Finslerian concepts extended to a probabilistic setting. The Laplace-Beltrami operator on TM results only when g_{ij} is independent of y, which is exactly the case when the drifts both vanish (i.e. Riemannian). That chapter also develops the diffusion theory on the *Indicatrix Bundle* of a Finsler manifold M, in keeping with the profound insights of E. Cartan, who viewed the subject as one taking place on a sphere bundle associated with the tangent bundle, but with spheres replaced by unit length vectors relative to the given Finslerian metric function. The main result is again that Finslerian Diffusion on the Indicatrix Bundle has the Laplace-Beltrami operator plus non-zero torsion-induced drift fields as generator. The results of Chapter 6 are of importance for *short-time asymptotics* of *Finslerian Diffusions* because they reduce the problem to a Riemannian one, albeit with torsion-induced drifts, which is already well-understood. The main obstacle would be the (ordinary) line integrals of these drifts, which enter the asymptotic kernel, [Mol75]. These would have to be numerically estimated is some cases.

A. Milgram and P. Rosenbloom (1950) studied harmonic forms on a compact, oriented, Riemannian manifold without boundary, using heat kernel methods from p.d.e. theory, [MR51]. In the Appendix A, we make a start on the Milgram-Rosenbloom program, but for the more general case of Finslerian manifolds. The first requirement is to construct a Finslerian Laplacian, Δ_{AZ}, acting on p-forms. This operator must live on M and not on TM, as do the generators of hv- and h-Brownian motions. The construction proceeds by first introducing a Finslerian version of a *random walk* and proving a Central Limit Theorem type result, which gives rise to our Laplacian Δ_{AZ}. This generalizes Mark Pinsky's theory of *isotropic transport* on a Riemannian manifold, [Pin78]. Then, following arguments of D. Bao and B. Lackey, [BL96], we show that each real de Rham cohomology class of M has a representative form which is annihilated by Δ_{AZ}. For more information on Finslerian Hodge Theory, the reader may consult *The Theory of Finslerian Laplacians and Applications*, eds. P. Antonelli and B. Lackey, to appear in Kluwer Academic Press. Appendix B contains a complete proof of the classification of *locally constant Berwald spaces of dimension two*. Such spaces play a major role in applications throughout the text.

Acknowledgements. Thanks to Paule Antonelli and Joseph Modayil for proof reading and to our friends Robert Elliott, Makoto Matsumoto, Robert Seymour, and Hideo Shimada who collaborated with us on various portions of the theory we here record. Financial support was provided in part by NSERC. The authors would also like to express their gratitude to Vivian Spak for her excellent typesetting.

Chapter 1

Finsler Spaces

1.1 The Tangent and Cotangent Bundle

The Roman letter M will denote a smooth (i.e. C^∞) *manifold* of dimension n. It will be assumed Hausdorff, connected and separable in the sense of having a countable base for its topology. The *smooth structure* is a family $\mathcal{F}(M)$ of coordinate charts $\{(U_\alpha, h_\alpha)\}$ which form an open covering of M, and for which all overlap maps

$$h_\beta \circ h_\alpha^{-1} : h_\alpha(U_\alpha \cap U_\beta) \to \mathbb{R}^n, \tag{1.1}$$

are smooth (i.e. C^∞), where each $h_\alpha : U_\alpha \to \mathbb{R}^n$, is a homeomorphism onto an open set of Euclidean n-space $\mathbb{R}^n$. It is also required that $\mathcal{F}$ is *maximal* with respect to the smoothness property of (1.1): if (U_γ, h_γ) has non-empty overlap with an element of $\mathcal{F}$, then it is itself in $\mathcal{F}$.

A *smooth* (i.e. C^∞) *map* $f : M \to N$, between smooth manifolds is a continuous map for which the collection of functions

$$k_\beta \circ f \circ h_\alpha^{-1} : h_\alpha(U_\alpha) \to k_\beta(V_\beta) \tag{1.2}$$

are smooth, when $(U_\alpha, h_\alpha) \in \mathcal{F}(M)$ and $(V_\beta, k_\beta) \in \mathcal{F}(N)$. Note that $h_\alpha(U_\alpha)$ is open in $\mathbb{R}^n$ and $k_\beta(V_\beta)$ is open in $\mathbb{R}^p$.

A *tangent vector* A to M at a point p in M is a map which assigns to each $(U_\alpha, h_\alpha) \in \mathcal{F}(M)$ for which $p \in U_\alpha$, an n-tuple A_α of real numbers, such that if (U_β, h_β) is another chart containing p, then

$$A_\beta = J(h_\beta \circ h_\alpha^{-1}) \cdot A_\alpha, \tag{1.3}$$

where $J(h_\beta \circ h_\alpha^{-1})$ denotes the Jacobian of the overlap map evaluated at $h_\alpha(p)$. Here, $A_\alpha = (A_\alpha^1, \ldots, A_\alpha^n)$ denotes the *contravariant components* of

A relative to the chart (U_α, h_α). The collection of tangent vectors at p is denoted by T_pM and constitutes a real vector space of dimension n. It is called the *tangent space* at p. Its operations are induced from componentwise addition and multiplication by a real scalar. None of the concepts described above depend on particular choices of coordinates.

For any tangent vector A at $p \in M$ and choice of chart around p, one can uniquely write

$$A_\alpha = A^i_\alpha \left(\frac{\partial}{\partial u^i}\right)_p. \tag{1.4}$$

(Summation on repeated upper and lower indices.) The dual vector space is the so-called *cotangent space* at p and is denoted T^*_pM. It has the *dual basis* du^i, $i = 1, \ldots, n$ given by

$$\langle \frac{\partial}{\partial u^i}, du^j \rangle_p = \delta^i_j \tag{1.5}$$

where $\langle , \rangle$ denotes evaluation of the linear functional du^j on the vector $(\partial/\partial u^i)$ and δ^i_j is the Kronecker delta or identity tensor. Elements of T^*_pM may be uniquely written as

$$B^\alpha = B^\alpha_j du^j. \tag{1.6}$$

The B^α_j are called the *covariant components* of the cotangent vector $B^\alpha = (B^\alpha_1, \ldots, B^\alpha_n)$ relative to the chart (U_α, h_α). The transformation from one chart to another is given by

$$B^\beta_j = J^i_j(h_\beta \circ h^{-1}_\alpha) \cdot B^\alpha_i \tag{1.7}$$

where $J^i_j(h_\beta \circ h^{-1}_\alpha)$ is the $(i,j)^{\text{th}}$ entry in the Jacobian of (1.3) above.

The collection of all tangent vectors to M is denoted $TM = \bigcup_{p\in M} T_pM$ while $T^*M = \bigcup_{p\in M} T^*_pM$ denotes the collection of all cotangent vectors to M. These are called, the *tangent bundle* and *cotangent bundle*, respectively. They are provided with a topology and smooth structure. First require the projection map $\pi : TM \to M$ defined by

$$\pi(A) = p \quad \text{iff} \quad A \in T_pM$$

to be continuous, while also requiring the map

$$\widetilde{h}_\alpha : \pi^{-1}(U_\alpha) \to \mathbb{R}^n \times \mathbb{R}^n \tag{1.8a}$$

given by

$$\left(h_\alpha(\pi(A)), A_\alpha\right) = \widetilde{h}_\alpha(A) \tag{1.8b}$$

to be a C^∞ diffeomorphism. The requisite overlap maps are given by

$$\widetilde{h}_\beta \circ \widetilde{h}_\alpha^{-1}(p, A) = \left(h_\beta \circ h_\alpha^{-1}(\overline{p}), J(h_\beta \circ h_\alpha^{-1}) \cdot A\right) \tag{1.9}$$

where $\overline{p} = h_\alpha(p)$ and the Jacobian is taken at $\overline{p}$. Since $h_\beta \circ h_\alpha^{-1}$ are C^∞ maps it follows that $\widetilde{h}_\beta \circ \widetilde{h}_\alpha^{-1}$ are also smooth. Hence, the charts $\left(\pi^{-1}(U_\alpha), \widetilde{h}_\alpha\right)$ generate a smooth structure on TM. It is easy to see that TM is Hausdorff, connected, separable and of dimension $2n$ and that the projection map π is smooth. A similar construction can be carried out for the cotangent bundle, T^*M, so that it too is a smooth $2n$-manifold. Both these bundles are examples of *fiber bundles* which we now briefly discuss. The basic reference is Steenrod's book, *The Topology of Fibre Bundles.*

1.2 Fiber Bundles

Recall that a topological transformation group G acting on a space F is, firstly, a topological group, meaning that G is a topological space for which the group operations $(g_1, g_2) \mapsto g_1 \cdot g_2$ and $g_1 \mapsto g_1^{-1}$ taking $G \times G \to G$ and $G \to G$, respectively, are continuous. Secondly, G acts on F (from the left), if $\left(g_1 \cdot (g_2 \cdot f)\right) = (g_1 \cdot g_2) \cdot f$ for all $g_1, g_2 \in G$ and $f \in F$. Here, $(g, f) \mapsto g \cdot f$ is the *action map* $G \times F \to F$. G is said to *act freely*, if $I \cdot f = f$ and $g \cdot f = f$ for *some* $f \in F$ implies $g = I$. On the other hand, if $I \cdot f = f$ and $g \cdot f = f$ for *all* $f \in F$ implies $g = I$, then G is said to *act effectively.* In this case, for any fixed g, the map $y \to g \cdot y$ is a homeomorphism whose inverse is $y \to g^{-1} \cdot y$. Therefore, G is isomorphic to a group of homeomorphisms of F.

Often G is a *Lie group*, meaning that G is a smooth manifold and that the two group operation maps are C^∞, as well. A nice example of a Lie group is, $GL(\mathbb{R}, n)$, the group of $n \times n$ real nonsingular matrices provided with the topology it receives from being viewed as an open subset of $\mathbb{R}^{n^2}$. It has two connected components, these being determined by the continuity of the real-valued determinant function defined on $GL(\mathbb{R}, n)$.

A *fiber bundle* is a 5-tuple (E, π, B, F, G) where E, B, F, G are topological spaces and $\pi : E \to B$ is a continuous map onto B with the following additional requirements. There is an open covering of B, $\{V_\alpha\}$ so that there are homeomorphisms

$$\varphi_\alpha : V_\alpha \times F \to \pi^{-1}(V_\alpha) \tag{1.10a}$$

for each V_α and

$$\pi\varphi_\alpha(p, f) = p \quad \text{for all} \quad p \in V_\alpha \tag{1.10b}$$

and $f \in F$.

Furthermore, there are continuous maps

$$h_{\alpha\beta} : V_\alpha \cap V_\beta \to G \tag{1.11a}$$

defined by homeomorphisms of F by

$$h_{\alpha\beta}(p) = \varphi_{\alpha,p}^{-1} \circ \varphi_{\beta,p}, \tag{1.11b}$$

with

$$\varphi_{\alpha,p}(f) = \varphi_\alpha(p, f), \tag{1.11c}$$

so that (1.11b) coincides with a *unique* element of G. The map $\varphi_{\alpha,p}$ of (1.11c) is a map of F into $\pi^{-1}(p)$. Thus, G is a topological transformation group acting (effectively) on F as a group of homeomorphisms.

The space E is called the *total space* of the fiber bundle, B, the *base space*, G, the *structural group* and, F, the *fiber space*. The map π is called the *projection map* of the fiber bundle. The special covering $\{V_\alpha\}$ is called the *trivializing cover* of the bundle. As stated above, the action of the structural group G on the fiber is always *effective*.

One can show that

$$\begin{aligned} h_{\gamma\alpha}h_{\alpha\beta} &= h_{\gamma\beta} \\ h_{\gamma\gamma} &= I \end{aligned} \tag{1.12}$$

holds on the triple overlap $V_\alpha \cap V_\beta \cap V_\gamma$. The properties (1.10a), (1.10b) and (1.12) characterize fiber bundles up to *bundle equivalence*. Two bundles with the same base, fiber and structural group are said to be bundle equivalent if their h-functions are conjugate in G. That is to say, if for each pair (α, β) there is a continuous map $\lambda : V_\alpha \cap V_\beta \to G$ such that

$$\widetilde{h}_{\alpha\beta}(p) = \lambda^{-1}(p) \cdot h_{\alpha\beta}(p)\lambda(p) \tag{1.13}$$

for all $p \in V_\alpha \cap V_\beta$. Of course, λ can be defined only when V_α and V_β have nonempty intersection.

If E, B, F, G are smooth manifolds and π is a smooth map of E onto B then λ and $h_{\alpha\beta}$ can be taken to be C^∞ maps, as well. In this case, G acts as a group of diffeomorphisms of the fiber manifold, F. Note that $\dim E = \dim B + \dim F$ and that π has *maximal rank* in the sense that its differential or Jacobian is a surjective linear map of T_pE to $T_{\pi(p)}B$ for each $p \in E$.

In the special case where E is exactly $B \times F$, the fiber bundle is said to be a *product bundle* or that it is *trivial*. This must always be the case if $G = \{I\}$, as can be seen directly from the definitions. However, it can be shown for B a Lie group, that the tangent bundle (a fiber bundle with $GL(\mathbf{R}, n)$ as structural group) is trivial. It follows that T^*G is trivial, as well, because G is *orientable*.

Here's why. First, a manifold M is called *orientable* if its tangent bundle TM is bundle equivalent to the same bundle, but with structural group $GL(\mathbf{R}, n)$ replaced by $SO(\mathbf{R}, n)$, the group of rotations. One says the structural group $GL(\mathbf{R}, n)$ has been *reduced* to $SO(\mathbf{R}, n)$. Second, if M is orientable, then TM and T^*M are bundle equivalent. Last, observe that any Lie group is orientable because its tangent bundle is actually reducible to $\{I\}$, because it is a product bundle.

An important class of fiber bundles is the so-called *principle bundles*, which have both fiber and group identical ($F = G$). Thus, G acts on itself effectively. In fact, this action is *free*. As an example let us consider $(S^3, \pi, S^2, S^1, S^1)$, the so-called Hopf Bundle, where complex numbers of modulus one (i.e. S^1) act by multiplication on the unit quaternions (i.e. S^3), the projection map π is just the quotient map of this action, which is without fixed points (i.e. free). The base space $B = S^2$, the unit 2-sphere, is just the quotient space. The fiber space is $S^1 = F$ and the structural group $G = S^1$ acts on F by left multiplication. On the other hand, (K, π, S^1, S^1, G) where K denotes the non-orientable, 2-dimensional manifold known as the Klein bottle, is the total space of a bundle with base space, S^1 and fiber space, S^1, but whose structural group, G, can not be taken to be, S^1. The reason being that, if it could, then the action of S^1 on $F = S^1 \subseteq K$ would be free with orbit space $B = S^1$ which is impossible because of non orientability of K. In fact, the group G can be reduced to $O(1)$, the orthogonal group of dimension one, i.e. Z_2, and the bundle is not principle.

In the next section we discuss the most important principle bundle in differential geometry, namely, the *frame bundle* of a smooth manifold. We follow A.I.M. rather closely, in this section.

1.3 Frame Bundles and Linear Connections

A *frame* z at $p \in M$ is a basis for T_pM, that is, a set $\{z_a\}$ $a = 1, \ldots, n$ of n linearly independent tangent vectors at p. We let L denote the set of all frames on M, and let $\pi : L \to M$ denote the projection $\pi(z) = p$, p being the origin of the frame z. The set of all frames with origin p is just $\pi^{-1}(p)$, the *fiber over p*. If (U_α, h_α) is a chart on M and $\partial/\partial x^i$, $i = 1, \ldots, n$, the basis for T_pM, $p \in U_\alpha$, then a frame $z = (x^i, z^i_a)$ evaluated at p is written $z^i_a(\partial/\partial x^i)_p$. The subset $\{\pi^{-1}(U_\alpha), (x^i, z^i_a)\} \subseteq L$ is a chart on the (n^2+n)-dimensional manifold L. The fiber is an n^2-dimensional submanifold with charts induced from those on L, i.e. (z^i_a). Define the *free right action* of $GL(\mathbf{R}, n) = G(n)$ on L by $\beta : L \times G(n) \to L$ with

$$\beta(z, g) = zg = (x^i, z^i_b g^b_a) \tag{1.14}$$

and using the summation convention on repeated upper and lower indices. For a fixed $g \in G(n)$, $\beta(z,g) = \beta_g(z) = zg \in L$, maps the fiber $\pi^{-1}(p)$ into itself. Therefore, M is the quotient space, $L/G(n)$, of the free right action β.

The 5-tuple $\big(L, \pi, M, G(n), G(n)\big)$ is a principle fiber bundle known as the *frame bundle on M*.

T_zL is the tangent space to L at z. There is a subvector space of tangent vectors along the fiber containing z, $L^v_z = \{X \in T_zL | d\pi(X) = 0\}$. This is just the kernel of the differential of π, the frame bundle projection. It is called the *vertical subspace* of T_zL.

Now fix $z \in L$ and note that ${}_z\beta : G(n) \to \pi^{-1}(p)$ given by the free right action β, has its differential, $d_z\beta$, mapping the tangent space, $T_eG(n)$, onto L^v_z. That is, the linear map $d_z\beta$ maps the tangent space of $G(n)$ at the identity e onto the vertical subspace at z. The former is just the *Lie algebra* $\widehat{G}(n)$ of $G(n)$ and consists of all $n \times n$ real matrices $(A^a_b) = A$. Then $d_z\beta(A) := Z(A)_z$. From (3.1) we see that

$$Z(A) = z^i_a A^a_b(\partial/\partial z^i_b), \quad A = (A^a_b), \quad z = (x^i, z^i_a) \in L, \tag{1.15}$$

since, $d_z\beta(\partial/\partial g^a_b)_e = z^i_a(\partial/\partial z^i_b)$. This *vertical vector field* on L is called the *fundamental vector field corresponding to A*.

What happens to $Z(A)$ under the free right action? Consider the mapping $h \in G(n) \to ghg^{-1} \in G(n)$ for fixed $g \in G(n)$. This is an inner automorphism of $G(n)$ and induces $\operatorname{ad}(g) : \widehat{G}(n) \to \widehat{G}(n)$ of the *Lie algebra*, defined by $\operatorname{ad}(g)A = \big(g^a_c A^c_d (g^{-1})^d_b\big)$ and is called the *adjoint representation of $g \in G(n)$*. Thus,

$$d_z\beta\big(Z(A)\big) = Z\big(\operatorname{ad}(g^{-1})A\big). \tag{1.16}$$

Now, just as there are *basic vector fields* on L there are *basic forms* on L. Let's briefly describe these. Define the map

$$\alpha : \pi^{-1}(p) \times V \to T_pM \tag{1.17a}$$

by

$$\alpha(z, v) = zv = z^i_a v^a (\partial/\partial x^i) \in T_pM \tag{1.17b}$$

with $z = (x^i, z^i_a)$, $v = v^a e_a$, where V is a real vector space of dimension n with basis (e_a), fixed once and for all. Now fix $z \in L$ and define the corresponding induced map from (3.4)

$$z^\alpha : v \in V \to zv \in T_pM \tag{1.18a}$$

and its inverse

$${}_z\alpha^{-1} : X^i \Big(\frac{\partial}{\partial x^i}\Big)_p \in T_pM \to z^{-1}v = (z^{-1})^a_i X^i e_a \in V. \tag{1.18b}$$

Here, $\big((z^{-1})^a_i\big)$ is the inverse matrix of (z^i_a). We define the *basic 1-form* by

$$\theta = {}_z\alpha^{-1} \circ d\pi = \{(z^{-1})^a_i dx^i\} e_a \tag{1.19}$$

from (1.18a), (1.18b) and composition with the differential of the frame bundle projection, $d\pi$. Note that since the vertical subspace L^v_z is the kernel of $d\pi$, we must have, $\theta(X) = 0$, for all vertical fields, i.e. $X \in L^r_z$. Furthermore, since

$${}_{(zg)}\alpha^{-1} = g^{-1}({}_z\alpha^{-1}),$$

we have

$$\theta \circ d\beta_g = g^{-1}\theta. \tag{1.20}$$

The relations (1.16) and (1.20), therefore settle the questions about action map β and its effects.

1.4 Tensor Fields

Let V^* denote the dual space of real vector space V of n dimensions. Thus, V^* is the set of all linear maps $V \to R$ with the basis e^a, $a = 1, \dots, n$, *dual* to that of V, e_b, i.e. such that, $e^a(e_b) = \delta^a_b$. Accordingly, $e^a(v) = v^a$ for vector $v = v^a e_a \in V$.

Now we have the left action of $G(n)$ on V defined by

$$\xi : (g, v) \in G(n) \times V \to gv = (g^a_b v^b) e_a \in V \tag{1.21}$$

with $g = (g^a_b)$ and $v = v^a e_a$. Thus, g acts on the basis (e_a) by $g(e_a) = (e_b g^b_a)$. This action is effective but not generally free.

Similarly to (1.21) we have

$$\xi^* : (g, v^*) \in G(n) \times V^* \to gv^* \in V^* \tag{1.22}$$

given by

$$gv^*(v) = v^*(g^{-1}v) \quad \text{for all} \quad v \in V\cdot$$

From (1.21) it follows that

$$gv^* = v_b(g^{-1})^b_a e^a,$$

for $v^* = v_a e^a$. The mapping α defined in (1.17a), (1.17b) can be defined for V^* by

$$\alpha^* : (z, v^*) \in \pi^{-1}(p) \times V^* \to zv^* = (z^{-1})^a_i v_a dx^i \in T^*_p M \tag{1.23a}$$

with $z = (x^i, z^i_a)$, $v^* = v_a e^a$ and $T^*_p M$ denoting the dual space of the tangent space $T_p M$. Note that α^* is characterized by

$$zv^*(zv) = v^*(v). \tag{1.23b}$$

Now we can define the *tensors* of *type* (r, s) as the set V^r_s of all multilinear mappings $V \underbrace{\times \cdots \times}_{s} V \times V^* \underbrace{\times \cdots \times}_{r} V^* \to \mathbb{R}$ provided with the basis $(e^{b_1 \ldots b_s}_{a_1 \ldots a_r})$ defined by

$$e^{b\ldots}_{a\ldots}(e^c, \ldots, e^d, \ldots) = \delta^c_a \ldots \delta^b_a \ldots \tag{1.24a}$$

This cumbersome notation is usually replaced with

$$\begin{aligned} V^r_s &= V \underbrace{\otimes \cdots \otimes}_{s} V \otimes V^* \underbrace{\otimes \cdots \otimes}_{r} V^* \quad \text{and} \\ e^{b\ldots}_{a\ldots} &= e_a \otimes \cdots \otimes e^b \otimes \ldots \end{aligned} \tag{1.24b}$$

The mapping ξ and ξ^* are also extended to (r, s)-tensors by

$$\xi^r_s : (g, w) \in G(n) \times V^r_s \to gw \in V^r_s \tag{1.25a}$$

given by

$$gw(v,\ldots,v^*,\ldots)=w(g^{-1}v,\ldots,g^{-1}v^*,\ldots). \qquad (1.25b)$$

Thus, for example,

$$gw=g_b^a g_e^b w_f^{de}(g^{-1})_c^f e_{ab}^c,$$

for $g=(g_b^a)$ and $w=w_c^{ab}e_{ab}^c\in V_2^1$.

If (u^i) are chart coordinates around point $p\in M$, a basis for T_pM is $(\partial/\partial u^i)_p$, while the dual basis for T_p^*M is $(du^i)_p$ according to (1.5). The tensor spaces $(T_pM)_s^r$ have bases $\partial/\partial u^i\otimes\cdots\otimes du^j\otimes\ldots$

By a *tensor field T on M of type (r,s)* we mean a V_s^r-valued function, $\widetilde{T}$, on the frame bundle L defined as follows:

$$\widetilde{T}:z\in L\to\widetilde{T}_z\in V_s^r, \qquad (1.26a)$$

with

$$\widetilde{T}_z(v,\ldots,v^*,\ldots)=T_p(zv,\ldots,zv^*,\ldots), \qquad (1.26b)$$

and $\pi(z)-p$, where, T_p is smooth as a function of u^i in a chart around p so that it sends $u\in M\to T_u\in(T_uM)_s^r$. If we let $T_{j\ldots}^{i\ldots}(u)$ be components of T in coordinates (u^i) then

$$\widetilde{T}_z=\widetilde{T}_{a\ldots}^{b\ldots}(z)e_{b\ldots}^{a\ldots},\ \widetilde{T}_{a\ldots}^{b\ldots}(z)=T_{i\ldots}^{j\ldots}(u)z_a^i\ldots(z^{-1})_j^b\ldots. \qquad (1.26c)$$

In fact,

$$\widetilde{T}_{zg}=(g^{-1})_a^b\ldots\widetilde{T}_{c\ldots}^{d\ldots}g_a^c\ldots e_{b\ldots}^{a\ldots}, \qquad (1.26d)$$

so that (1.25b) yields

$$\widetilde{T}\circ\beta_g=g^{-1}\widetilde{T}. \qquad (1.27)$$

We shall make good use of tensors from now on in our text. The first topic to consider is of profound importance. It is the theory of Linear Connections on Manifolds.

1.5 Linear Connections

In a smooth manifold N a *smooth map* $D : u \in N \to V_u \in T_uN$ where V_u is a subspace of the tangent vector spaces T_uN is called a *distribution* in N. There is a special distribution called the *vertical distribution* $L^v : z \in L \to L^v_z \in T_zL$ in the total space L of the frame bundle of M. It is identical to the kernel of the basic 1-form θ and is spanned at each point $z \in L$ by the fundamental vectors $Z(A)$.

A distribution $\Gamma : z \in L \to \Gamma_z \in T_zL$ in the total space L of the frame bundle is called a *linear connection* in L (or on M), if

1) $T_zL = \Gamma_z \oplus L^v_z$

2) $d\beta_g(\Gamma_z) = \Gamma_{zg}, \quad g \in G(n)$.

The subspace Γ_z of T_zL is called the *horizontal subspace* and $X \in \Gamma_z$ are called *horizontal vector fields.* A *vector field* on M induces by scalar multiplication a smooth distribution of 1-dimensional subspaces, since it is a smooth map $u \in M \to T_u \in (T_uM)^0_1$. Likewise, a 1-*form* induces via scalar multiplication a 1-dimensional distribution on T^*M, since it is a smooth map $u \in M \to T^*_u \in (T_uM)^1_0$. The reader should note that on many occasions throughout this text, distributions, tensor fields, connections etc. will be only locally defined.

By the *lift* with respect to Γ we mean a map $\ell_z : T_pM \to \Gamma_z, \quad \pi(z) = p$ such that

$$d\beta_g \circ \ell_z = \ell_{zg}. \tag{1.28}$$

Note that Γ is invariant under the free right action, according to (2) above, and is a complement of the vertical distribution.

For a given linear connection Γ we can define a $\widehat{G}(n)$-valued differential 1-form, W, called the *connection form of* Γ, by

1) $W(Z(A)) = A, \quad A \in \widehat{G}(n)$

2) $W(\Gamma_z) = 0$.

Thus, (3.3) and the definition of W imply

$$W \circ d\beta_g = ad(g^{-1})W. \tag{1.29}$$

There is therefore, a one-to-one correspondence between linear connections Γ and connection forms W.

Consider the canonical coordinates in the frame bundle, namely, (u^i, z^i_a) : $W_z = (W(z)^a_b)$ where

$$W(z)^a_b := (z^{-1})^a_i(dz^i_b + z^j_b\Gamma^i_{jk}du^k), \tag{1.30}$$

where Γ^i_{jk} depend only on u^i. These h^3 functions are the so-called *connection coefficients* of Γ. Now, (2) in the definition of W leads to

$$\ell_z(X) = X^i\Big(\frac{\partial}{\partial u^i} - z^j_a\Gamma^k_{ji}\frac{\partial}{\partial z^k_a}\Big), \tag{1.31}$$

where $X = X^i(\partial/\partial u^i)_p \in T_pM$. Thus, to each choice of $v \in V$, the standard fiber ($\mathbb{R}^n$) of TM, we obtain a *horizontal vector field*, $\mathbb{B}(v)$, given by

$$\mathbb{B}(v)_z = \ell_z(zv) = z^i_a v^a\Big(\frac{\partial}{\partial u^i} - z^j_b\Gamma^k_{ji}\frac{\partial}{\partial z^k_b}\Big) \tag{1.32}$$

with $z = (u^i, z^i_a)$, $v = v^a e_a$.

This $\mathbb{B}(v)$ is called the *basic vector field* corresponding to $v \in V$. Thus, (1.28) gives

$$d\beta_g(\mathbb{B}(v)_z) = \mathbb{B}(g^{-1}v)_{zg}. \tag{1.33}$$

To sum up, a linear connection Γ yields a connection form, W, supplementary to the basic form θ, and the basic vector fields $\mathbb{B}(v)$ supplementary to the fundamental vector field $Z(A)$, satisfying

$$\begin{array}{lll} 1) & W(Z(A)) = A, & W(\mathbb{B}(v)) = 0 \\ 2) & \theta(Z(A)) = 0, & \theta(\mathbb{B}(v)) = v. \end{array} \tag{1.34}$$

Now consider the V^r_s-valued function $\widetilde{T}$ on L corresponding to a tensor field T of type (r,s) as in (1.26a) - (1.26d). The fundamental vector field $Z(A)$ acts on $\widetilde{T}$ so that

$$Z(A)(\widetilde{T}) = -A\cdot\widetilde{T}, \tag{1.35}$$

as follows from (1.26b) and (1.15). For instance, if $(r,s) = (1,2)$ and $w \in V^1_2$, $A\cdot w$ is just $A^a_d w^d_{bc} - w^a_{dc}A^d_b - w^a_{bd}A^d_c$ and $\partial(z^{-1})^a_i/\partial z^j_b = -(z^{-1})^a_j(z^{-1})^b_i$.

In a similar manner we define *covariant derivative* of *tensor field* T of type (r,s) to be ∇T, where on the frame bundle L,

$$\widetilde{\nabla T}(v) := \mathbb{B}(v)(\widetilde{T}). \tag{1.36}$$

Thus, ∇T is a tensor of type $(r, s+1)$. For instance, for $(r,s) = (1,1)$ we must have

$$\mathbf{B}(v_2)(\widetilde{T}(v_1, v^*)) = \widetilde{\nabla T}(v_1, v_2, v^*), \quad v_1, v_2 \in V, \quad v^* \in V^*. \tag{1.37}$$

Then from (1.32) and (1.26d) we obtain the components of the covariant derivative ∇T as

$$(\nabla T)^i_{jk} = \partial_k T^i_j + T^r_j \Gamma^i_{rk} - T^i_r \Gamma^r_{jk}, \tag{1.38}$$

which is also expressed as $T^i_{j;k}$ in classical notation.

1.6 Torsion and Curvature of a Linear Connection

Recall that in the Lie algebra $\widetilde{G}(n)$ there is defined the Lie bracket, $[A_1, A_2] := A_1A_2 - A_2A_1$. It is easy to see, then, that

$$[Z(A_1), Z(A_2)] = Z([A_1, A_2]) \tag{1.39a}$$

and

$$[Z(A), \mathbf{B}(v)] = \mathbf{B}(A \cdot v) \tag{1.39b}$$

where $A \cdot v$ is defined as in the example (1.35).

Now consider the Lie bracket of $\mathbf{B}(v_1)$ and $\mathbf{B}(v_2)$ and note

$$[\mathbf{B}(v_1), \mathbf{B}(v_2)] = \mathbf{B}(\widetilde{T}(v_1, v_2)) + Z(\widetilde{\mathbf{R}}(v_1, v_2)) \tag{1.40}$$

where $\widetilde{T}(v_1, v_2) \in V$ and $\widetilde{\mathbf{R}}(v_1, v_2) \in \widehat{G}(n)$ which can be identified with a tensor of type $(1,1)$. Since $\widetilde{R} \circ \beta_g = g^{-1}\widetilde{R}$ and $\widetilde{T} \circ \beta_g = g^{-1}\widetilde{T}$ from (1.16) and (1.34) we are lead to tensor fields $\mathbf{R}$ of (1,3)-type and $\mathbb{T}(v_1, v_2)$ of (1,2)-type called the Γ-*curvature tensor* and Γ-*torsion tensor*, respectively. Use of (1.32) give us for curvature

(1) $\mathbf{R}^i_{hjk} = \partial_k \Gamma^i_{hj} + \Gamma^r_{hj}\Gamma^i_{rk} - (j/k)$,

where (j/k) denotes the terms obtained from those preceeding by interchanging indices j and k.

Likewise, for the torsion we have

(2) $T^i_{jk} = \Gamma^i_{jk} - (j/k)$.

Note that both (1) and (2) are local expressions, in a fixed chart on M.

Using the classical notation we can observe readily the *Ricci identity* (for contravariant vector fields)

$$A^i{}_{;h;k} - (h/k) = \mathbf{R}^i_{rhk} A^r - A^i{}_{;r} T^r_{hk}. \tag{1.41}$$

Similar expressions hold for other (r, s)-tensors. Using the *Jacobi identity* for the Lie bracket of vector fields, applied to basic vector fields, we get

$$\big[[\mathbf{B}(v_1), \mathbf{B}(v_2)], \mathbf{B}(v_3)\big] + (1,2,3) = 0, \tag{1.42}$$

where $(1, 2, 3)$ denotes the terms obtained from the first one by cyclic permutation. Substitution from (1.40) into this leads to the *Bianchi identities* by splitting into horizontal and vertical parts, namely,

$$\mathbf{T}^h_{ij;k} + \mathbf{T}^h_{ir} \mathbf{T}^r_{jk} - \mathbf{R}^h_{ijk} + (i, j, k) = 0 \tag{1.43}$$

and

$$\mathbf{R}^\ell_{hij;k} + \mathbf{R}^\ell_{hir} \mathbf{T}^r_{jk} + (i, j, k) = 0. \tag{1.44}$$

1.7 Parallelism

If $C : [0, 1] \to L$ is a smooth curve on the frame bundle, then $\pi \circ C = \overline{C}$, is a smooth curve on M which is said to be *covered* by C. If the tangent vectors of C are horizontal relative to a linear connection Γ, then C is called the *horizontal lift* of $\overline{C}$. In a local coordinate chart (u^i) on M and the canonical coordinates (u^i, z^i_a) in L, $\overline{C}$ is simply $(u^i(t))$ whereas, C is $(u^i(t), z^i_a(t))$ and (1.31) shows

$$du^i/dt = X^i, \quad dz^i_a/dt = -z^j_a \Gamma^i_{jk}\big(u(t)\big) X^k. \tag{1.45}$$

A *frame field* $z(t)$ along $\overline{C}$ on M is called *parallel along* $\overline{C}$, if the smooth curve $t \to z(t)$ is horizontal. A vector field $v(t)$ along $\overline{C}$ is called *parallel along* $\overline{C}$, if $v(t)$ *has constant components in a parallel frame field along* $\overline{C}$. It follows that (1.45) gives a parallel frame field, and if $v(t) = v^i(t)(\partial/\partial u^i)$, $v^i(t) = z^i_a(t) v^a$, with constant v^a, then transvection with v^a gives

$$dv^i/dt + v^j \Gamma^i_{jk}\big(u(t)\big) du^k/dt = 0, \tag{1.46}$$

as the equations for a *parallel vector field*. Generally, for a *contravariant vector field* A^i along $\overline{C}$ in M, the *absolute differential* relative to Γ is defined as

$$DA^i := dA^i + A^j \Gamma^i_{jk}\big(u(t)\big) du^k, \tag{1.47}$$

in a local chart (u^i). *Infinitesimal parallel displacement* of a vector field A requires $DA^i = 0$. Likewise, the absolute differential of a *covariant vector field*, A_i, is

$$DA_i := dA_i - A_i^j \Gamma_{jk}\big(u(t)\big) du^k. \tag{1.48}$$

1.8 The Levi-Cività Connection on a Riemannian Manifold

Suppose M is provided with a *positive-definite* tensor g of type $(0,2)$, i.e. for X, Y vector fields

$$g(X,Y) = g(Y,X), \qquad g(X,X) \geq 0, \tag{1.49a}$$

with $g(X,X) = 0$, if and only if $X = 0$. In a local chart on M, (U,h) we can write

$$g(X,Y) = g(\xi^i \partial_i, \eta^j \partial_j) = g_{ij}\xi^i \eta^j, \tag{1.49b}$$

where $g_{ij} = g(\partial_i, \partial_j)$ and $\partial_i = \partial/\partial u^i$. This so-called *Riemannian metric tensor*, g_{ij}, enables one to define *arc lengths* along any smooth curve, $u^i = u^i(t)$, by

$$s = \int_{t_1}^{t_2} \sqrt{g_{ij}\big(u(t)\big) \frac{du^i}{dt} \frac{du^j}{dt}}\, dt, \tag{1.50}$$

where du/dt is the *velocity vector field* of the curve. One is also able to define *angles* between intersecting curves in M at the point p of intersection, as the angle θ defined by,

$$\cos\theta = \frac{g(X_p, Y_p)}{\sqrt{g_{rs}X_p^r X_p^s} \cdot \sqrt{g_{\ell m} Y_p^\ell Y_p^m}}\,, \tag{1.51}$$

where X_p and Y_p are the velocity fields of the two curves at intersection p.

Another important idea is that the Riemannian metric tensor allows us to *raise and lower indices of tensors*. For instance,

$$\mathbb{R}_{mjk\ell} = g_{mi}\mathbb{R}^i_{jk\ell} \tag{1.52}$$

define the *purely covariant components* of $\mathbb{R}$ and

$$\mathbb{R} := g^{ij}R_{ij} := g^{ij}\mathbb{R}^k_{ijk} \tag{1.53}$$

defines the *scalar curvature*, $\mathbf{R}$, and the *Ricci curvature*, $\mathbf{R}_{ij}$. Of course,

$$g^{ij} g_{j\ell} = \delta^i_\ell$$

so that (g^{ij}) in a particular local chart is the matrix inverse of $(g_{j\ell})$ in that chart.

But, the role of the metric tensor is even more profound than the above, because it can be used to construct a linear connection Γ which satisfies, via the associated Γ-covariant derivative,

$$g_{ij;k} := \partial_k g_{ij} - g_{rj}\Gamma^r_{ik} - g_{ri}\Gamma^r_{jk} = 0. \tag{1.54}$$

This is the *Ricci Condition*. We say also that a connection Γ satisfying (1.54) is *metrical*. The *fundamental theorem of Riemannian geometry* is simply that there is exactly one linear metrical connection with $T^i_{jk} = 0$, on a Riemannian Manifold, (M, g). A connection with zero torsion is called *symmetric*. This unique symmetric metrical connection is called the Levi-Cività connection, after its founder. It has the local coefficients (i.e. Γ's)

$$\gamma^i_{jk} = \frac{1}{2} g^{i\ell}(\partial_k g_{\ell j} + \partial_j g_{\ell k} - \partial_\ell g_{ik}). \tag{1.55}$$

More generally, if we relax the symmetry condition on the connection Γ one can show by direct calculation that

$$\begin{aligned} \Gamma_{ijk} &= g_{jr}\Gamma^r_{ik} = \gamma_{ijk} + A_{ijk}, \\ A_{ijk} &= \frac{1}{2}(T_{ijk} - T_{jki} + T_{kij}), \\ T_{ijk} &= g_{jr}T^r_{ik}, \end{aligned} \tag{1.56}$$

where $\gamma_{ijk} = g_{jr}\gamma^r_{ik}$ is given by (1.55).

Thus, if the *torsion tensor* T^i_{jk} is specified a priori there is a *unique* linear connection Γ which is metrical, i.e. so that (1.54) holds.

1.9 Geodesics, Stability and the Orthonormal Frame Bundle

It is often convenient to use a Riemannian metric on a given manifold M, as they always exist. Therefore, there are a lot of linear connections, too, by the results in Section 8 above. It is clear parallel displacement via the Levi-Cività connection does indeed preserve the length of vectors in a Riemannian manifold, because of its metricity. Similarly, the angle θ between a pair of tangent vectors at p is preserved; as follows from (1.51). Therefore, since $g_{ij}\frac{dx^i}{ds}\frac{dx^j}{ds}=1$ is preserved under parallel transport, the covariant derivative of $dx^i/ds=\dot{x}^i$, vanishes i.e. $D(\dot{x}^i)=0$ or

$$\frac{d^2x^i}{ds^2}+\Gamma^i_{jk}\frac{dx^j}{ds}\frac{dx^k}{ds}=0. \tag{1.57}$$

The trajectories or solutions of (1.57) are called *geodesics.* In fact, (1.57) are Euler-Lagrange equations for the (regular) variational problem

$$\frac{d}{d\gamma}\int_\gamma ds=0, \tag{1.58}$$

with fixed endpoints and ds given by (1.50). The positive definiteness of g_{ij} ensures that geodesics actually minimize length, locally. Let $x^i(s;\varepsilon)$ be a family of geodesics with initial conditions $x^i(0;\varepsilon)$ and $\dot{x}^i(0;\varepsilon)=\xi^i(0;\varepsilon)$, $i=1,2,\ldots,n$. The $x^i(s;\varepsilon)$ are C^∞ relative to ε. For $|\varepsilon|\leq 1$ and each i,

$$x^i(s;\varepsilon)=x^i(s)+\varepsilon V^i(s)+\varepsilon^2(\cdots) \tag{1.59}$$

from which it follows that

$$\Gamma^i_{jk}(x(s;\varepsilon))=\Gamma^i_{jk}(x(s))+\partial_\ell\Gamma^i_{jk}\varepsilon V^\ell(s)+\varepsilon^2(\cdots). \tag{1.60}$$

Substituting into (1.59) and (1.60) into (1.57) we obtain

$$\varepsilon\frac{d^2V^i}{ds^2}=-\varepsilon\partial_\ell\Gamma^i_{jk}V^\ell\frac{dx^j}{ds}\frac{dx^k}{ds}-2\varepsilon\Gamma^i_{jk}\frac{dx^k}{ds}\frac{dV^j}{ds}+\varepsilon^2(\cdots). \tag{1.61}$$

Dividing by ε and letting $\varepsilon\to 0$, we obtain the *deviation equations* or *Jacobi field* (V^ℓ) *equations,*

$$\frac{d^2V^i}{ds^2}+\partial_\ell\Gamma^i_{jk}V^\ell\frac{dx^j}{ds}\frac{dx^k}{ds}+2\Gamma^i_{jk}\frac{dx^k}{ds}\frac{dV^j}{ds}=0. \tag{1.62a}$$

Direct calculations now show (1.62a) can be written

$$\frac{D^2V^i}{ds^2}+\mathbb{R}^i_{jk\ell}\frac{dx^k}{ds}\frac{dx^j}{ds}V^\ell=0, \tag{1.62b}$$

which is the *invariant form* of the deviation equations. Here, $\mathbb{R}^i_{jk\ell}$ are just the coefficients given in (1) of Section 6. If V^i is a solution of (1.62a), (1.62b) so that initially $V^i(s_0)$ is orthogonal to the velocity field of geodesic γ, then this is true for all $s \geq s_0$.

Stability of geodesics, in the *sense of Liapunov*, can be decided by $\mathbb{R}$, $\mathbb{R}_{ij}$, and $\mathbb{R}^i_{jh\ell}$. For example, if $\mathbb{R} \leq 0$, then geodesics are (locally) *unstable* while $\mathbb{R} > 0$ ensures *stability*, for the case $n = 2$. For higher dimensions one must require the so-called *sectional curvatures*, defined below, all be positive for stability. Just one of these being non-positive yields Liapunov instability. Let $X_r := \xi^k_{(r)}(\partial/\partial u^k)$ be n *mutually orthogonal unit vectors* at $p \in (U, h)$ on M. Thus,

$$g_{ij}(p)\xi^i_{(r)}\xi^j_{(s)} = \delta_{rs}, \tag{1.63}$$

and $\{X_1, \ldots, X_n\}$ is called an *orthonormal frame* at p. The *sectional curvatures* K are determined by the set of all pairs (X_r, X_s), by

$$K(r, s) = -\mathbb{R}_{ijk\ell}\xi^i_{(r)}\xi^j_{(s)}\xi^k_{(r)}\xi^\ell_{(s)}. \tag{1.64}$$

Just as the frame bundle L over M has been defined above using arbitrary frames, one can restrict the frames to *orthonormal* frames defined by (1.63) for a Riemannian manifold (M, g). However, $GL(\mathbb{R}, n)$ would be replaced by $O(\mathbb{R}, n)$, the n-dimensional orthogonal group, in the construction. Alternatively, using the fact that $O(\mathbb{R}, n)$ is a compact Lie subgroup, which is also a deformation retract of $GL(\mathbb{R}, n)$, the bundle reduction theorem in Steenrod's book (ibid.) shows that the *orthonormal frame bundle*, thus constructed, is *bundle equivalent* (in $GL(\mathbb{R}, n)$) *to* L. We shall denote OM, for the *orthonormal frame bundle* over M in the rest of this text.

1.10 Finsler Space and Metric

Let M be an n-dimensional smooth, i.e. C^∞, manifold. Then TM will denote the tangent bundle over M with projection $\pi_T : TM \to M$, so that $\pi_T(y) = x$ if $y \in TM_x$, TM_x being the tangent space at $x \in M$.

Given any smooth curve $[a, b] \ni t \mapsto \sigma(t) \in M$ with velocity $\dot{\sigma}(t) \in TM_{\sigma(t)}$, we define the *length* of σ by

$$I(\sigma) = \int_a^b L(\sigma(t), \dot{\sigma}(t))\, dt,$$

where $L(x, y)$ is a scalar function defined for any $x \in M$ and $y \in TM_x$, which can thus be regarded as a scalar field on TM, $L : TM \to \mathbb{R}$. It is assumed that $L(x, y)$ satisfies the conditions below.

(L1) *$L(x,y)$ is positively homogeneous of degree one in y, that is,*

$$L(x,ky) = kL(x,y)$$

for any $x \in M$, $y \in TM_x$, and $k > 0$.

We observe that condition (L1) ensures that $I(X)$ is independent of the parametrization of X, as long as the change of parametrization preserves the orientation of the curve.

(L2) *$L(x,y)$ is smooth at every $x \in X$ and $y \in TM_x$ such that $y \neq 0$.*

Note that if $L(x,y)$ is also differentiable at $y = 0$, then the homogeneity condition (L1) implies that $L(x,y)$ is linear in y.

Next, let us consider a local chart $(x^i) : M \supset U \to \mathbf{R}^n$ on a coordinate neighbourhood U in M. For any $x \in U$ and $y \in TM_x$, we can write $y = y^i \partial_i$, where $\partial_i = \partial/\partial x^i$. Then $(x^i, y^i) : TM \supset \pi_T^{-1}U \to \mathbf{R}^{2n}$ is a local chart on TM. We call (x^i, y^i) the *induced coordinates* on TM.

Since, as a consequence of (L1), $L^2(x,y)$ is homogeneous of degree two in y, it can be written as $L^2(x,y) = g_{ij}(x,y)y^i y^j$ in terms of the induced coordinates, $g_{ij}(x,y)$ being homogeneous of degree zero in y. If $g_{ij}(x,y)$ is assumed to be symmetric in i,j, then it is uniquely defined by this formula. By Euler's theorem on homogeneous functions, it is seen that

$$g_{ij}(x,y) = \frac{1}{2}\frac{\partial^2 L^2(x,y)}{\partial y^i \partial y^j}. \tag{1.65}$$

It follows immediately that g can be defined globally as a function from the so-called *slit tangent bundle* $\mathring{T}M$, that is, TM with the zero section removed, to the bundle $T_2^0 M$ of tensors of rank $(0,2)$ over M such that $g(x,y) \in T_2^0 M_x$ for any $x \in M$ and $0 \neq y \in TM_x$.

In addition to (L1) and (L2), we assume that

(L3) *$g_{ij}(x,y)$ is a non-degenerate positive definite quadratic form for any $x \in M$ and $y \in TM_x$ such that $y \neq 0$.*

The latter assumption is clearly a generalization of the positive definiteness condition for the metric tensor in Riemannian geometry. It proves crucial for the existence of a Brownian motion.

Definition 1.1. A *Finsler space* is a finite-dimensional smooth manifold M equipped with $L : TM \to \mathbf{R}$ such that conditions (L1), (L2), (L3) are satisfied. The scalar function L is then called the *Finsler metric function* and g defined by (1.65) is called the *Finsler metric tensor.*

1.11 Finsler Tensor Fields

We have seen above that the Finsler metric tensor g is a smooth function from the slit tangent bundle $\mathring{T}M$ (that is, TM with the zero section removed) to the bundle T^0_2M of tensors of rank $(0,2)$ on M such that $g(x,y) \in T^0_2M_x$ for any $x \in M$ and $0 \neq y \in TM_x$. Similarly, the Finsler metric function L is a smooth scalar function from $\mathring{T}M$ to $\mathbb{R}$. This observation gives rise to the following general definition.

Definition 1.2. A *Finsler tensor field* A of rank (m,n) on $\mathring{T}M$ is a smooth function from the slit bundle $\mathring{T}M$ to the tensor bundle T^m_nM such that $A(x,y) \in T^m_nM_x$ for any $x \in M$ and $0 \neq y \in TM_x$, that is, the following diagram commutes:

$$\begin{array}{ccc} & A & \\ \mathring{T}M & \rightarrow & T^m_nM \\ \pi_1 \searrow & & \swarrow \pi_2 \\ & M & \end{array}$$

Here π_1 and π_2 are the projections of the corresponding bundles over M.

The above definition should be compared with the standard definition of a tensor field A on M as a smooth function from M to T^m_nM such that $A(x) \in T^m_nM_x$ for any $x \in M$, that is, a smooth section of the tensor bundle T^m_nM:

$$\begin{array}{ccc} & A & \\ M & \leftrightarrows & T^m_nM \\ & \pi_2 & \end{array}$$

By $\mathcal{F}^m_n(\mathring{T}M)$ we shall denote the set of Finsler tensor fields of rank (m,n) on $\mathring{T}M$. We shall write simply $\mathcal{F} = \mathcal{F}^1_0(\mathring{T}M)$ for Finsler vector fields. We shall denote by $\mathcal{X}^m_n(M)$ the set of standard tensor fields of rank (m,n) on a manifold M and write simply $\mathcal{X} = \mathcal{X}^1_0(M)$ for standard vector fields on M.

It follows that $g : \mathring{T}M \rightarrow T^0_2M$ defined by (1.65) is a Finsler tensor field in $\mathcal{F}^0_2(\mathring{T}M)$. It is called the *Finsler metric tensor.* By condition (L3) of the previous section, the matrix (g_{ij}) representing g in local coordinates is non-singular. Its inverse matrix (g^{ij}) determines another Finsler tensor field, which belongs to $\mathcal{F}^2_0(\mathring{T}M)$. One can use (g^{ij}) and (g_{ij}) to raise and lower the indices of any Finsler tensor field in the standard way.

As another example, we consider the so-called *supporting element* $\eta \in \mathcal{F}$, a Finsler vector field defined by $\eta(x,y) = y$ for any $x \in M$ and $0 \neq y \in TM_x$.

1.12 Nonlinear Connections

Let (x^i) be local coordinates on M with the associated induced coordinates (x^i, y^i) on TM. We shall consider the tangent bundle $T\mathring{T}M$ over $\mathring{T}M$ referred to as the *double tangent bundle.* $T\mathring{T}M_{x,y}$ will denote the tangent space to $\mathring{T}M$ at (x,y). We shall write $\partial_i = \partial/\partial x^i$ meaning the canonical vector fields on M, and ${}^T\partial_i = \partial/\partial x^i$ and ${}^T\dot{\partial}_i = \partial/\partial y^i$ for the induced vector fields on $\mathring{T}M$ in local coordinates.

Assigning $\partial_i \mapsto {}^T\dot{\partial}_i$, one can readily obtain a globally defined linear mapping v from the space $\mathcal{F}$ of Finsler vector fields on $\mathring{T}M$ to the space $\mathcal{X}$ of standard vector fields on $\mathring{T}M$, defined locally by $A^i\partial_i \mapsto A^i\,{}^T\dot{\partial}_i$, A^i being the local components of a Finsler vector field $A \in \mathcal{F}$. The vectors ${}^T\dot{\partial}_i$, $i = 1, \ldots, n$ span an n-dimensional subspace $V\mathring{T}M_{x,y}$ in $T\mathring{T}M_{x,y}$ for every (x,y) in the corresponding coordinate neighbourhood, the subspace being independent of the choice of local coordinates. The subspaces $V\mathring{T}M_{x,y}$ form the so-called *vertical distribution* $V\mathring{T}M$ over $\mathring{T}M$. The linear mapping $v : \mathcal{F} \to \mathcal{X}$ is called the *vertical lift* over $\mathring{T}M$.

Talking of ${}^T\dot{\partial}_i$, let us also mention that, applying this operator to any Finsler field A of rank (m,n), we can get a new Finsler field of rank $(m, n+1)$ with components ${}^T\dot{\partial}_i A^{j_1,\ldots,j_m}_{k_1,\ldots,k_n}$.

On the other hand, trying to use ${}^T\partial_i$ instead of ${}^T\dot{\partial}_i$ as above, presents difficulties. The mapping $\partial_i \mapsto {}^T\partial_i$ cannot be extended to a global linear mapping from $\mathcal{F}$ to $\mathcal{X}$. The subspaces spanned by ${}^T\partial_i$, $i = 1 \ldots, n$ depend on the choice of local coordinates in M. And, lastly, ${}^T\partial_i$ applied to a Finsler tensor field does not, in general, produce a Finsler tensor field, not even if the given tensor field is a scalar Finsler field. The concept of a nonlinear connection as described below provides a solution to these problems.

For any $x \in M$ and $0 \neq y \in TM_x$, $V\mathring{T}M_{x,y}$ is an n-dimensional subspace of $T\mathring{T}M_{x,y}$. We can write $T\mathring{T}M_{x,y} = H\mathring{T}M_{x,y} \oplus V\mathring{T}M_{x,y}$, where $H\mathring{T}M_{x,y}$ is an arbitrarily chosen n-dimensional subspace and $\oplus$ denotes the direct sum. The distribution $H\mathring{T}M$ over $\mathring{T}M$ formed by all the subspaces $H\mathring{T}M_{x,y}$ is called a *horizontal distribution.* We write $T\mathring{T}M = H\mathring{T}M \oplus V\mathring{T}M$.

Definition 1.3. By a *nonlinear connection* over $\mathring{T}M$ we mean any smooth horizontal distribution $H\mathring{T}M$ over $\mathring{T}M$.

For any (x,y) in a local coordinate neighbourhood on $\mathring{T}M$, there is exactly one vector ${}^T\delta_i$ in $H\mathring{T}M_{x,y}$ that projects onto ${}^T\partial_i$. It can be written as

$${}^T\delta_i = {}^T\partial_i - N^j_i(x,y)\,{}^T\dot{\partial}_j.$$

We call $N_i^j(x,y)$ the *nonlinear connection coefficients.* The vectors ${}^T\delta_i$ span the horizontal space $H\mathring{T}M_{x,y}$. We say that ${}^T\delta_i$ and ${}^T\dot{\partial}_i$ form the *adapted frame.*

We can now assign $\partial_i \mapsto {}^T\delta_i$. This can be extended to a global linear mapping $h : \mathcal{F} \to \mathcal{X}$ called the *horizontal lift* over $\mathring{T}M$.

Corresponding to any nonlinear connection with coefficients $N_j^i(x,y)$ is the notion of parallelism along a curve $\sigma : [a,b] \ni t \mapsto x(t) \in M$. We say that a curve $\tilde{\sigma} : [a,b] \ni t \mapsto (x(t), y(t)) \in \mathring{T}M$ with $0 \neq y(t) \in TM_{x(t)}$ is a *horizontal lift* of σ if $\tilde{\sigma}$ is tangent to the horizontal distribution $H\mathring{T}M$, that is, $d\tilde{\sigma}(t)/dt \in H\mathring{T}M_{\tilde{\sigma}(t)}$ for all $t \in [a,b]$. In local coordinates this can be written as

$$\frac{dy^i}{dt} = -N_j^i(x,y)\frac{dx^j}{dt}.$$

1.13 Affine Connections on the Finsler Bundle

A modern approach to Finsler connections is to represent them as connections on the *Finsler bundle* FM, a principal bundle over $\mathring{T}M$ whose total space consists of elements (x,y,z), where $x \in M$, $0 \neq y \in TM_x$, and $z = (z_1, \ldots, z_n)$ are linear frames in TM_x, with the projection $\pi_F : FM \ni (x,y,z) \mapsto (x,y) \in \mathring{T}M$. The structure group is $GL(n)$, the group of non-singular linear transformations $g : \mathbb{R}^n \to \mathbb{R}^n$ with composition as the group operation. The right action of $GL(n)$ on FM is $R_g(x,y,z) = (x,y,zg)$, where $(zg)_j = z_i g_j^i$ with (g_j^i) being the matrix of $g \in GL(n)$ in the standard basis of $\mathbb{R}^n$. Then $\mathring{T}M$ can be identified with the quotient space $FM/GL(n)$.

Given a local chart $(x^i) : M \supset U \to \mathbb{R}^n$ on M, we define a local chart $(x^i, y^i, z_j^i) : FM \supset \pi_F^{-1}\pi_T^{-1}U \to \mathbb{R}^{2n+n^2}$ by writing $y = y^i\,\partial_i$ and $z_j = z_j^i\,\partial_i$. We call (x^i, y^i, z_j^i) the *induced coordinates* on FM. In this way the total space FM can be equipped with the structure of a smooth manifold. The right action of the structure group on FM is then a smooth mapping.

The Finsler bundle FM on which to introduce affine connections is used with a view to developing the concept of parallelism for Finsler vector fields. The slit bundle $\mathring{T}M$ is the base space of FM because it is the domain of Finsler fields. The fibres are linear frames in the tangent spaces to M because TM is the range of Finsler vector fields. Once an affine connection on FM is specified, it determines linear frames parallel along a curve σ in $\mathring{T}M$. Then Finsler vector fields parallel along σ, and in fact Finsler tensor fields of any rank parallel along σ, are defined as those whose components are constant in any parallel frame along σ.

We consider the tangent bundle TFM over FM. Given local coordinates (x^i) on M, we set ${}^F\partial_i = \partial/\partial x^i$, ${}^F\dot{\partial}_i = \partial/\partial y^i$, and ${}^F\partial_i^j = \partial/\partial z_j^i$, where (x^i, y^i, z_j^i) are the induced coordinates on FM. The vectors ${}^F\partial_i^j$ span an n^2-dimensional subspace VFM_u in TFM_u for every $u = (x, y, z)$ in the corresponding coordinate neighbourhood. The subspaces VFM_u are independent of the choice of local coordinates and form the so-called *vertical distribution* VFM over FM. For any $u \in FM$, we can choose a $2n$-dimensional subspace HFM_u such that $TFM_u = HFM_u \oplus VFM_u$. By a *horizontal distribution* HFM over FM we mean a distribution formed by such horizontal subspaces HFM_u. We write $TFM = HFM \oplus VFM$.

Definition 1.4. By an *affine connection* over FM we mean any smooth horizontal distribution HFM over FM that is invariant under the right action of $GL(n)$, i.e.,

$$TR_g(HFM) = HFM \qquad \text{for any} \quad g \in GL(n),$$

$TR_g : TFM \to TFM$ being the differential of $R_g : FM \to FM$.

For any $u = (x, y, z)$ in a local coordinate neighbourhood on FM, there is exactly one vector $l({}^T\partial_i)$ in HFM_u that projects onto ${}^F\partial_i$ and exactly one vector $l({}^T\dot{\partial}_i)$ that projects onto ${}^F\dot{\partial}_i$. These vectors can be written as

$$l({}^T\partial_i) = {}^F\partial_i - \Gamma_{ij}^k(x,y) z_l^j \, {}^F\partial_k^l, \tag{1.66}$$

$$l({}^T\dot{\partial}_i) = {}^F\dot{\partial}_i - C_{ij}^k(x,y) z_l^j \, {}^F\partial_k^l. \tag{1.67}$$

We call $\Gamma_{ij}^k(x,y)$ and $C_{ij}^k(x,y)$ the *coefficients* of the affine connection. That they are independent of z follows from the invariance of HFM under the right action of $GL(n)$. The vectors $l({}^T\partial_i)$ and $l({}^T\dot{\partial}_i)$, where $i = 1, \ldots, n$, span the horizontal space HFM_u. On assigning ${}^T\partial_i \mapsto l({}^T\partial_i)$ and ${}^T\dot{\partial}_i \mapsto l({}^T\dot{\partial}_i)$, we can extend l to a global linear mapping $l : \mathcal{X} \to \mathcal{X}$ called the *horizontal lift* of the affine connection HFM.

An affine connection on FM with coefficients $\Gamma_{jk}^i(x,y)$ and $C_{jk}^i(x,y)$ gives rise to the notion of parallelism along a curve $\sigma : [a,b] \ni t \mapsto (x(t), y(t)) \in TM$ with $x(t) \in M$ and $0 \neq y(t) \in TM_{x(t)}$. A curve $\tilde{\sigma} : [a,b] \ni t \mapsto (x(t), y(t), z(t)) \in FM$ with $z(t)$ being a linear frame in $TM_{x(t)}$ is said to be a *horizontal lift* of σ if $\tilde{\sigma}$ is tangent to the horizontal distribution HFM, that is, $d\tilde{\sigma}(t)/dt \in HFM_{\tilde{\sigma}(t)}$ for every $t \in [a,b]$. In local coordinates this can be written as

$$\frac{dz_l^i}{dt} = -\Gamma_{jk}^i(x,y) z_l^k \frac{dx^j}{dt} - C_{jk}^i(x,y) z_l^k \frac{dy^j}{dt}.$$

We also say that the linear frame $z(t)$ is *parallel* along σ. Of a Finsler tensor field $A \in \mathcal{F}_n^m(\mathring{T}M)$ we say that it is *parallel along* σ if the components of $A(\sigma(t))$ with respect to a linear frame $z(t)$ that is parallel along σ are independent of t. For instance, a Finsler vector field $A \in \mathcal{F}$ is a parallel along σ if $A(\sigma(t)) = \alpha^i z_i(t)$ with $\alpha^i = \text{const}$. We define the so-called *pull-back transformation* along σ by

$$T_{s,t} : TM_{x(s)} \ni \alpha^i z_i(s) \mapsto \alpha^i z_i(t) \in TM_{x(t)}, \qquad \alpha^i \in \mathbf{R}$$

for any $s, t \in [a, b]$, where $z(t)$ is a linear frame parallel to σ. The above definitions are independent of the choice of the linear frame $z(t)$ parallel along σ. A Finsler vector field $A \in \mathcal{F}$ is a parallel along σ if $T_{s,t}A(\sigma(s)) = A(\sigma(t))$ for any $s, t \in [a, b]$. In local coordinates the condition for $A \in \mathcal{F}$ to be parallel along $\sigma(t) = (x(t), y(t))$ reads

$$\frac{dA^i(x,y)}{dt} = -\Gamma^i_{jk}(x,y)A^k(x,y)\frac{dx^j}{dt} - C^i_{jk}(x,y)A^k(x,y)\frac{dy^j}{dt}. \tag{1.68}$$

An affine connection on FM determines the *covariant derivative* $\nabla_Z A \in \mathcal{F}$ of a Finsler vector field $A \in \mathcal{F}$ relative to a vector field $Z \in \mathcal{X}$. This covariant derivative can be defined by taking any integral curve σ of Z, that is, any curve $\sigma : [a, b] \to \mathring{T}M$ such that $d\sigma/dt = Z \circ \sigma$, and setting

$$(\nabla_Z A)(\sigma(t)) = \lim_{h \to 0} \frac{T_{t+h,t}A(\sigma(t+h)) - A(\sigma(t))}{h} \tag{1.69}$$

for any $t \in (a, b)$. This defines $\nabla_Z A$ at every point (x, y) of $\mathring{T}M$, since there is always an integral curve of Z passing through (x, y). In local coordinates the expression for $\nabla_Z A$ reads

$$\nabla_Z A = \left[Z^j \left({}^T\partial_j A^i + \Gamma^i_{jk} A^k \right) + Z^{(j)} \left({}^T\dot{\partial}_j A^i + C^i_{jk} A^k \right) \right] \partial_i, \tag{1.70}$$

where $Z = Z^i\, {}^T\partial_i + Z^{(i)}\, {}^T\dot{\partial}_i$ and $A = A^i\, \partial_i$. The definition of $\nabla_Z A$ can be extended in the standard way to Finsler tensor fields $A \in \mathcal{F}_n^m(\mathring{T}M)$ of any rank (m, n); cf. (1.73) and (1.74) in the next section.

It follows that the condition for a Finsler tensor field $A \in \mathcal{F}_n^m(\mathring{T}M)$ to be parallel along a smooth curve σ on $\mathring{T}M$ can be expressed as $(\nabla_Z A) \circ \sigma = 0$, where $Z \in \mathcal{X}$ is any vector field on $\mathring{T}M$ such that σ is an integral curve of Z.

1.14 Finsler Connections

Formula (1.69) cannot be applied directly to define the covariant derivative of a Finsler vector field $A \in \mathcal{F}$ with respect to another Finsler vector field $X \in \mathcal{F}$, for there is no integral curve σ on $\mathring{T}M$ associated directly to X. One can, however, use a nonlinear connection over $\mathring{T}M$ to lift X to a vector field in $\mathcal{X}$, and then, given an affine connection over FM, use the covariant derivative (1.69) of A with respect to the lifted field. This leads to the following definition, which combines the connections of the preceding two sections.

Definition 1.5. A *Finsler connection* on M is a pair consisting of a nonlinear connection $H\mathring{T}M$ over $\mathring{T}M$ and an affine connection HFM over FM.

Consider a Finsler connection on M whose nonlinear connection $H\mathring{T}M$ has coefficients $N_i^j(x,y)$ and whose affine connection HFM has coefficients $\Gamma_{ij}^k(x,y)$ and $C_{ij}^k(x,y)$ in local coordinates. Composing the horizontal and vertical lifts h and v of the nonlinear connection $H\mathring{T}M$ with the horizontal lift l of HFM, we obtain the linear mappings $l \circ h, l \circ v : \mathcal{F} \to \mathcal{X}$, which act as follows when applied to ∂_i:

$$\begin{aligned} l \circ h(\partial_i) &= l(^T\delta_i) = {}^F\delta_i - F_{ij}^k(x,y) z_l^j \, {}^F\dot{\partial}_k^l, \\ l \circ v(\partial_i) &= l(^T\dot{\partial}_i) = {}^F\dot{\partial}_i - C_{ij}^k(x,y) z_l^j \, {}^F\dot{\partial}_k^l, \\ {}^F\delta_i &= {}^F\partial_i - N_i^j(x,y) {}^F\dot{\partial}_j, \end{aligned} \tag{1.71}$$

where

$$F_{ij}^k(x,y) = \Gamma_{ij}^k(x,y) - N_i^l(x,y) C_{lj}^k(x,y).$$

We call $N_i^j, F_{ij}^k, C_{ij}^k$ the *coefficients* of the Finsler connection. It is customary to refer to a Finsler connection simply by specifying the triple $(N_i^j, F_{ij}^k, C_{ij}^k)$.

Since either h or v can be used to lift a Finsler vector field $X \in \mathcal{F}$ to a vector field in $\mathcal{X}$, there are two corresponding covariant derivatives of a Finsler vector field $A \in \mathcal{F}$ with respect to X defined by

$$\nabla_X^h A = \nabla_{h(X)} A, \qquad \nabla_X^v A = \nabla_{v(X)} A. \tag{1.72}$$

These are called the *horizontal* and *vertical covariant derivatives*, respectively. The covariant derivative on the right-hand sides in (1.72) is defined by (1.69). It follows from (1.70) and (1.71) that

$$\begin{aligned} \nabla_X^h A &= X^j \left({}^T\delta_j A^i + F_{jk}^i A^k \right) \partial_i, \\ \nabla_X^v A &= X^j \left({}^T\dot{\partial}_j A^i + C_{jk}^i A^k \right) \partial_i \end{aligned}$$

in local coordinates.

The definition of $\nabla^h_X A$ and $\nabla^v_X A$ can be extended in the standard way to Finsler tensor fields $A \in \mathcal{F}^m_n(\mathring{T}M)$ of arbitrary rank (m, n). Denoting the components of A in local coordinates by $A^{i_1 \dots i_m}_{j_1 \dots j_n}$, we set

$$A^{i_1 \dots i_m}_{j_1 \dots j_n \, |k} = {}^T\delta_k A^{i_1 \dots i_m}_{j_1 \dots j_n} + \sum_{\mu=1}^{m} A^{i_1 \dots i_{\mu-1} l i_{\mu+1} \dots i_m}_{j_1 \dots j_n} F^{i_\mu}_{kl} - \sum_{\nu=1}^{n} A^{i_1 \dots i_m}_{j_1 \dots j_{\nu-1} l j_{\nu+1} \dots j_n} F^l_{kj_\nu}, \tag{1.73}$$

$$A^{i_1 \dots i_m}_{j_1 \dots j_n}\Big|_k = {}^T\dot{\partial}_k A^{i_1 \dots i_m}_{j_1 \dots j_n} + \sum_{\mu=1}^{m} A^{i_1 \dots i_{\mu-1} l i_{\mu+1} \dots i_m}_{j_1 \dots j_n} C^{i_\mu}_{kl} - \sum_{\nu=1}^{n} A^{i_1 \dots i_m}_{j_1 \dots j_{\nu-1} l j_{\nu+1} \dots j_n} C^l_{kj_\nu}. \tag{1.74}$$

Then, for any $X \in \mathcal{F}$, the components of $\nabla^h_X A$ and $\nabla^v_X A$ are $X^k A^{i_1 \dots i_m}_{j_1 \dots j_n \, |k}$ and $X^k A_{j_1 \dots j_n}\big|^{i_1 \dots i_m}_k$, respectively.

Let $\sigma : [a, b] \in t \to (x(t), y(t)) \in \mathring{T}M$ with $x(t) \in M$ and $0 \neq y \in TM_{x(t)}$ be a curve in $\mathring{T}M$. To express the condition for a Finsler tensor field $A \in \mathcal{F}^m_n(\mathring{T}M)$ to be parallel along σ in terms of the Finsler connection, we take a vector field $Z \in \mathcal{X}$ for which σ is an integral curve, that is, $Z \circ \sigma = d\sigma/dt$, and consider the Finsler vector fields $X, Y \in \mathcal{F}$ uniquely determined by the relation $Z = h(X) + v(Y)$. It follows that

$$X^i(x, y) = \frac{dx^i}{dt}, \qquad Y^i(x, y) = \frac{\delta y^i}{dt}, \tag{1.75}$$

where

$$\frac{\delta y^i}{dt} = \frac{dy^i}{dt} + N^i_j(x, y)\frac{dx^j}{dt}. \tag{1.76}$$

It also follows that $\nabla_Z A = \nabla^h_X A + \nabla^v_Y A$. Then the condition $(\nabla_Z A) \circ \sigma = 0$ for A to be parallel along σ can be written as

$$(\nabla^h_X A + \nabla^v_Y A) \circ \sigma = 0. \tag{1.77}$$

In particular, for a Finsler vector field $A \in \mathcal{F}$, the latter condition expressed in local coordinates reads

$$\frac{dA^i(x, y)}{dt} = -F^i_{jk}(x, y) A^k(x, y) \frac{dx^j}{dt} - C^i_{jk}(x, y) A^k(x, y) \frac{\delta y^j}{dt},$$

which is equivalent to (1.68) via (1.71) and (1.76).

As an example we consider the horizontal covariant derivative of the supporting element $\eta \in \mathcal{F}$, a Finsler vector field defined in Sect. 1.11 by $\eta(x,y) = y$ for any $x \in M$ and $0 \neq y \in TM_x$. There is a Finsler tensor field $D \in \mathcal{F}^1_1(\mathring{T}M)$ called the *deflection tensor* such that $\nabla^h_X \eta = D(X) = D^i_j X^j \partial_i$ for any $X \in \mathcal{F}$. It follows that

$$D^i_j(x,y) = y^i{}_{|j} = F^i_{jk}(x,y)y^k - N^i_j(x,y)$$

in local coordinates. A Finsler connection is said to be *deflection-free* if $D = 0$.

From now on, when there is no risk of confusing vectors form different tangent bundles, we shall suppress the indices ^{T} and ^{F} attached to ∂_i, δ_i, $\dot{\partial}_i$, and $\partial^j_{\mathring{i}}$.

1.15 Torsions and Curvatures of a Finsler Connection

We consider a Finsler connection with coefficients $N^i_j, F^i_{jk}, C^i_{jk}$ and designate the horizontal and vertical covariant derivatives by $_|$ and $|$ as in (1.73) and (1.74), respectively. For any Finsler vector field $A \in \mathcal{F}$, one can easily verify the Ricci identities

$$A^i{}_{|j|k} - A^i{}_{|k|j} = A^m R^i_{mkj} - A^i{}_{|m} T^m_{kj} - A^i|_m R^m_{kj}, \qquad (1.78)$$

$$A^i{}_{|j}|_k - A^i|_{k|j} = A^m P^i_{mkj} - A^i{}_{|m} C^m_{kj} - A^i|_m P^m_{kj}, \qquad (1.79)$$

$$A^i|_j|_k - A^i|_k|_j = A^m S^i_{mkj} - A^i|_m S^m_{kj}. \qquad (1.80)$$

The five Finsler tensor fields $T, R, C, P, S \in \mathcal{F}^1_2(\mathring{T}M)$ with components T^i_{jk}, R^i_{jk}, C^i_{jk}, P^i_{jk}, S^i_{jk} are called the *torsions* of the Finsler connection. We have $T(X,Y) = T^i_{jk} X^j Y^k \partial_i$ for any $X, Y \in \mathcal{F}$, and similar formulae for $R(X,Y)$, $C(X,Y)$, $P(X,Y)$, and $S(X,Y)$. The components C^i_{jk} of one of the torsion fields turn out to be the same as the connection coefficients C^i_{jk}. The components of the other four torsions are expressed in terms of the connection coefficients as follows:

$$\begin{aligned} T^i_{jk} &= F^i_{jk} - F^i_{kj}, \\ R^i_{jk} &= \delta_j N^i_k - \delta_k N^i_j, \\ S^i_{jk} &= C^i_{jk} - C^i_{kj}, \\ P^i_{jk} &= \dot{\partial}_j N^i_k - F^i_{jk}. \end{aligned}$$

The three Finsler tensor fields $R, P, S \in \mathcal{F}^1_3(\mathring{T}M)$ with components R^i_{jkl}, P^i_{jkl}, S^i_{jkl} are called the *curvatures* of the Finsler connection. We have $R(A,X,Y) = R^i_{jkl}A^jX^kY^l\partial_i$ for any $A,X,Y \in \mathcal{F}$, and similar formulae for $P(A,X,Y)$ and $S(A,X,Y)$. The expressions for the curvature fields in terms of the connection coefficients read

$$\begin{aligned} R^i_{jkl} &= \delta_k F^i_{lj} - \delta_l F^i_{kj} + F^i_{km}F^m_{lj} - F^i_{lm}F^m_{kj} + (\dot\partial_k N^m_l - \dot\partial_l N^m_k)C^i_{mj}, \\ P^i_{jkl} &= \dot\partial_k F^i_{lj} - \delta_l C^i_{kj} + C^i_{km}F^m_{lj} - F^i_{lm}C^m_{kj} + \dot\partial_k N^m_l C^i_{mj}, \\ S^i_{jkl} &= \dot\partial_k C^i_{lj} - \dot\partial_l C^i_{kj} + C^i_{km}C^m_{lj} - C^i_{lm}C^m_{kj}. \end{aligned} \tag{1.81}$$

It is customary to use the same letters R, P, S to designate both the curvature fields and the corresponding three torsion fields. This abuse of notation is unlikely to lead to ambiguity, as the curvature and torsion fields have different rank.

Below we present an index-free description of the torsion and curvature fields of a Finsler connection. For any $Z, W \in \mathcal{X}$ such that $Z = hX + vY$ and $W = hU + vV$ with $X, Y, U, V \in \mathcal{F}$, we set

$$\tilde\nabla_Z W = h\nabla^h_X U + h\nabla^v_Y U + v\nabla^h_X V + v\nabla^v_Y V \in \mathcal{X}.$$

We call $\tilde\nabla$ the *lifted covariant derivative.* We define the *lifted torsion* and *curvature tensors* $\tilde T \in \mathcal{X}^1_2(\mathring{T}M)$ and $\tilde R \in \mathcal{X}^1_3(\mathring{T}M)$ by

$$\tilde T(Z,W) = \tilde\nabla_Z W - \tilde\nabla_W Z - [Z,W], \tag{1.82}$$

$$\tilde R(Q,Z,W) = \tilde\nabla_Z\tilde\nabla_W Q - \tilde\nabla_W\tilde\nabla_Z Q - \tilde\nabla_{[Z,W]}Q \tag{1.83}$$

for any $Q, Z, W \in \mathcal{X}$. Please note that formulae (1.82) and (1.82) cannot be written directly for Finsler vector fields without further elaboration of the concept of the Lie bracket $[\,\cdot\,,\cdot\,]$ for such fields.

The one-to-one correspondence between vector fields $A \in \mathcal{X}$ and pairs of Finsler vector fields $X, Y \in \mathcal{F}$ expressed by the relation $A = hX + vY$ induces a one-to-one correspondence between tensor fields in $\mathcal{X}^m_n(\mathring{T}M)$ and 2^{m+n}-tuples of Finsler tensor fields in $\mathcal{F}^m_n(\mathring{T}M)$. In the case of $\tilde T \in \mathcal{X}^1_2(\mathring{T}M)$ the corresponding eight Finsler fields in $\mathcal{F}^1_2(\mathring{T}M)$ are determined by

$$\begin{aligned} \tilde T(hX,hY) &= hT(X,Y) + vR(X,Y), \\ -\tilde T(hX,vY) &= hC(Y,X) + vP(Y,X), \\ \tilde T(vX,hY) &= hC(X,Y) + vP(X,Y), \\ \tilde T(vX,vY) &= vS(X,Y) \end{aligned} \tag{1.84}$$

for any $X, Y \in \mathcal{F}$. It is seen that one of the eight Finsler fields on the right-hand side of (1.84) is equal to zero, and the remaining ones are expressed in terms of the Finsler torsion fields $T, R, C, P, S \in \mathcal{F}^1_2(\overset{\circ}{T}M)$. In the case of the lifted curvature field $\tilde{R} \in \mathcal{X}^1_3(\overset{\circ}{T}M)$, the corresponding 16 Finsler fields in $\mathcal{F}^1_3(\overset{\circ}{T}M)$ are determined by

$$\begin{aligned} \tilde{R}(hA, hX, hY) &= hR(A, X, Y), & \tilde{R}(vA, hX, hY) &= vR(A, X, Y), \\ -\tilde{R}(hA, hX, vY) &= hP(A, Y, X), & -\tilde{R}(vA, hX, vY) &= vP(A, Y, X), \\ \tilde{R}(hA, vX, hY) &= hP(A, X, Y), & \tilde{R}(vA, vX, hY) &= vP(A, X, Y), \\ \tilde{R}(hA, vX, vY) &= hS(A, X, Y), & \tilde{R}(vA, vX, vY) &= vS(A, X, Y) \end{aligned} \tag{1.85}$$

for any $A, X, Y \in \mathcal{F}$. Eight of these 16 Finsler fields are seen to be zeros, while the remaining eight are expressed in terms of the Finsler curvature fields $R, S, T \in \mathcal{F}^1_3(\overset{\circ}{T}M)$. Relations (1.84) and (1.85) can serve as definitions of the Finsler torsion and curvature fields appearing on the right-hand sides.

1.16 Metrical Finsler Connections. The Cartan Connection

The theory of Finsler connections presented in the preceding sections is developed with no reference to the metric of a Finsler manifold. Here we shall be concerned with Finsler connections compatible with the metric in the sense of the definition below.

Definition 1.6. A *metrical Finsler connection* is one for which the Finsler metric tensor g is parallel along any smooth curve σ on $\overset{\circ}{T}M$.

Given a smooth curve σ on $\overset{\circ}{T}M$, we can choose two Finsler vector fields $X, Y \in \mathcal{F}$ such that σ is an integral curve for $Z = hX + vY \in \mathcal{X}$. According to (1.77), g is parallel along σ if $\left(\nabla^h_X g + \nabla^v_Y g\right) \circ \sigma = 0$. Since σ is arbitrary, it follows that

$$\nabla^h_X g = 0, \qquad \nabla^v_X g = 0 \tag{1.86}$$

for any $X \in \mathcal{F}$. Conditions (1.86) are necessary and sufficient for a Finsler connection to be metrical. It is said that a Finsler connection is *h-metrical* if $\nabla^h_X g = 0$ and *v-metrical* if $\nabla^v_X g = 0$ for any $X \in \mathcal{F}$.

We consider the submanifold OM of FM that consists of points $u = (x, y, z) \in FM$ such that $x \in M$, $0 \neq y \in TM_x$, and $z = (z_1, \dots, z_n)$ is a linear frame in TM_x orthonormal relative to $g(x, y)$. This submanifold can be equipped with the structure of a principal bundle over $\mathring{T}M$ with projection $\pi_O : OM \ni (x, y, z) \mapsto (x, y) \in \mathring{T}M$ and structure group $O(n)$ consisting of all orthogonal transformations $g : \mathbf{R}^n \to \mathbf{R}^n$ with composition as the group operation. The right action of $O(n)$ on OM is $R_g(x, y, z) = (x, y, zg)$, where $(zg)_j = z_i g^i_j$, (g^i_j) being the matrix of $g \in O(n)$ in the standard basis of $\mathbf{R}^n$. Then $\mathring{T}M$ can be identified with the quotient space $OM/O(n)$. We shall call OM the *orthonormal Finsler bundle.*

Another equivalent condition for a Finsler connection to be metrical is that

$$HFM_u \subset TOM_u \qquad \text{for every } u \in OM, \tag{1.87}$$

where HFM is the corresponding horizontal distribution over FM and TOM denotes the tangent bundle to OM.

To demonstrate the sufficiency of (1.86), we take a smooth curve $\sigma : [a, b] \to \mathring{T}M$ and consider a horizontal lift $\tilde{\sigma} : [a, b] \to FM$ of σ, so that $\tilde{\sigma}(t) \in FM_{\sigma(t)}$ and $d\tilde{\sigma}(t)/dt \in HFM_{\tilde{\sigma}(t)}$ for all $t \in [a, b]$, as described in Sect. 1.13. This determines $\tilde{\sigma}$ to within the initial condition $\tilde{\sigma}(a) \in FM_{\sigma(a)}$. We can choose $\tilde{\sigma}(a) \in OM_{\sigma(a)}$. Then, by (1.87), it follows that $\tilde{\sigma}(t) \in OM$ for all $t \in [a, b]$. Hence, for all $t \in [a, b]$, the components of $g(\sigma(t))$ in the frame $\tilde{\sigma}(t)$ are equal to δ_{ij}, i.e., constant. Thus, by definition, g is parallel along σ, which proves the sufficiency of (1.87).

To verify the necessity of (1.87), we fix any $u \in OM$ and take an arbitrary vector field $Z \in \mathcal{X}$. Let $\sigma : [a, b] \to \mathring{T}M$ be an integral curve of Z for which $u \in OM_{\sigma(a)}$, and let $\tilde{\sigma} : [a, b] \to FM$ be the horizontal lift of σ with $\tilde{\sigma}(a) = u \in OM$. Since g is parallel along σ, it follows that $\tilde{\sigma}(t) \in OM$ for all $t \in [a, b]$. Thus $(lZ)_u = d\tilde{\sigma}(a)/dt \in TOM_u$, where $l : \mathcal{X} \to \mathcal{X}$ is the horizontal lift defined by (1.66) and (1.67). It follows that (1.87) holds true, because every vector in HFM_u is of the form $(lZ)_u$ for some $Z \in \mathcal{X}$.

We note that condition (1.87) enables one to regard the affine part of a metrical Finsler connection as an affine connection over the orthonormal Finsler bundle OM, rather than over FM, the horizontal and vertical distributions HOM and VOM over OM being defined by $HOM_u = HFM_u$ and $VOM_u = VFM_u \cap TOM_u$ for every $u \in OM$. We then have $TOM = HOM \oplus VOM$. It is easily verified that HOM is invariant under the right action of $O(n)$. We call the resulting connection the *induced connection over* OM. It will be referred to frequently when dealing with metric Finsler connections later on.

It is also true that any affine (i.e. invariant under the right action of $O(n)$) connection HOM over OM can be extended to a metrical affine (i.e., satisfying (1.87) and invariant under the right action of $GL(n)$) connection HFM over FM. The reader may wish to check the details.

The principal example of a metrical Finsler connection has been introduced by Cartan and called by him the Euclidean connection. Nowadays it is customary to refer to it as the Cartan connection. It plays a similar role in Finsler geometry as the Levi-Cività connection in Riemannian geometry. Below the Cartan connection is defined by a system of axioms due to Matsumoto.

Definition 1.7. The *Cartan connection* is the unique Finsler connection that

(C1) is h-metrical, $\nabla^h_X g = 0 \;\; \forall\, X \in \mathcal{F}$;

(C2) is v-metrical, $\nabla^v_X g = 0 \;\; \forall\, X \in \mathcal{F}$;

(C3) its torsion tensor T vanishes identically, $T = 0$;

(C4) its torsion tensor S vanishes identically, $S = 0$;

(C5) its deflection tensor D vanishes identically, $D = 0$.

We shall denote the coefficients of the Cartan connection by $\mathbf{N}^i_j, \mathbf{F}^i_{jk}, \mathbf{C}^i_{jk}$. These coefficients can be computed from (C1)–(C5). Let us write (C1) and (C2) as

$$\delta_k g_{ij} = \mathbf{F}^l_{ki} g_{lj} + \mathbf{F}^l_{kj} g_{li}, \qquad \dot{\partial}_k g_{ij} = \mathbf{C}^l_{ki} g_{lj} + \mathbf{C}^l_{kj} g_{li}. \tag{1.88}$$

Since, by (C3) and (C4), $\mathbf{F}^i_{jk} = \mathbf{F}^i_{kj}$ and $\mathbf{C}^i_{jk} = \mathbf{C}^i_{kj}$, we can find in much the same way as in the case of the Levi-Cività connection that

$$\mathbf{F}^i_{jk} = \frac{1}{2} g^{il} \left(\delta_j g_{lk} + \delta_k g_{lj} - \delta_l g_{jk} \right), \tag{1.89}$$

$$\mathbf{C}^i_{jk} = \frac{1}{2} g^{il} \left(\dot{\partial}_j g_{lk} + \dot{\partial}_k g_{lj} - \dot{\partial}_l g_{jk} \right). \tag{1.90}$$

But $g_{ij} = \dot{\partial}_i \dot{\partial}_j L^2/2$, and so (1.90) can be written simply as

$$\mathbf{C}^i_{jk} = \frac{1}{2} g^{il} \dot{\partial}_l g_{jk}. \tag{1.91}$$

Since $g_{ij}(x, y)$ is homogeneous of degree zero in y, it follows from (1.91) that

$$\mathbf{C}^i_{jk} y^k = 0 \tag{1.92}$$

by Euler's theorem on homogeneous functions. To determine $\mathbf{F}^i_{jk}$ and $\mathbf{N}^i_j$, we write (C4) as $\mathbf{N}^i_j = \mathbf{F}^i_{jk}y^k$. Note that the formula (1.89) for $\mathbf{F}^i_{jk}$ involves $\mathbf{N}^i_j$ via $\delta_j = \partial_j - \mathbf{N}^i_j \dot{\partial}_i$. We substitute this expression in (1.89), multiply by y^k, and use (1.92) to get

$$\mathbf{N}^i_j = \mathbf{F}^i_{jk}y^k = \gamma^i_{jk}y^k - \mathbf{C}^i_{jk}\mathbf{N}^k_l y^l, \tag{1.93}$$

where

$$\gamma^i_{jk} = \frac{1}{2}g^{il}\left(\partial_j g_{lk} + \partial_k g_{lj} - \partial_l g_{jk}\right). \tag{1.94}$$

On multiplying (1.93) by y^j and using (1.92) once more, we obtain $\mathbf{N}^i_j y^j = \gamma^i_{jk}y^j y^k$. Thus, we finally find from (1.93) that

$$\mathbf{N}^i_j = \gamma^i_{jk}y^k - \mathbf{C}^i_{jk}\gamma^k_{lm}y^l y^m. \tag{1.95}$$

The coefficients $\mathbf{N}^i_j$, $\mathbf{F}^i_{jk}$, $\mathbf{C}^i_{jk}$ are thus completely determined by (1.89), (1.91), and (1.95) in terms of the metric tensor g_{ij} and its derivatives. This also proves the uniqueness of the Cartan connection.

Chapter 2

Introduction to Stochastic Calculus on Manifolds

2.1 Preliminaries

The standard setting for stochastic calculus is a *probability space* $(\Omega, \mathcal{F}, P)$ equipped with a *filtration* $(\mathcal{F}_t)_{t\geq 0}$. Here Ω is a set, $\mathcal{F}$ is a *σ-field* on Ω, that is, a family of subsets of Ω satisfying the conditions

(F1) $\Omega \in \mathcal{F}$,

(F2) *if* $A \in \mathcal{F}$, *then* $\Omega \setminus A \in \mathcal{F}$,

(F3) *if* $A_1, A_2, \ldots \in \mathcal{F}$, *then* $\bigcup_{i=1}^{\infty} A_i \in \mathcal{F}$,

and $P : \mathcal{F} \to \mathbf{R}$ is a *probability measure* on $\mathcal{F}$ defined by the conditions

(P1) $P(A) \geq 0$ *for every* $A \in \mathcal{F}$,

(P2) $P(\Omega) = 1$,

(P3) *if* $A_1, A_2, \ldots \in \mathcal{F}$ *are disjoint sets, then* $P(\bigcup_{i=1}^{\infty} A_i) = \sum_{i=1}^{\infty} P(A_i)$.

We shall always assume that the probability space is *complete*, that is, if $A \subset B \in \mathcal{F}$ and $P(B) = 0$, then $A \in \mathcal{F}$.

The sets $A \in \mathcal{F}$ are called *events* and the elements $\omega \in \Omega$ are called *elementary events*. If a property holds for all $\omega \in A \setminus B$, where $A, B \in \mathcal{F}$ and $P(B) = 0$, it is said to hold *P-almost surely* (*P*-a.s.) *on* A. In particular, if $A = \Omega$, then we simply say that the property holds *P*-a.s.

A *random variable* with values in a topological space M is a function $\xi : \Omega \to M$ that is $\mathcal{F}$-measurable, that is, such that $\xi^{-1}(B) \in \mathcal{F}$ for every $B \in \mathcal{B}$, where $\mathcal{B}$ denotes the σ-field of Borel-measurable subsets of M. Given a family of random variables $(\xi_i)_{i \in I}$ with values in M, we denote by $\sigma(\xi_i, i \in I)$ the σ-field *generated* by the family. It is defined as the smallest σ-field that contains all the sets $\xi_i^{-1}(B)$, where $i \in I$ and $B \in \mathcal{B}$. It follows that $\sigma(\xi_i, i \in I) \subset \mathcal{F}$. We shall write $\sigma(\xi)$ for the σ-field generated by a single random variable ξ.

We say that σ-fields $\mathcal{F}_1, \ldots, \mathcal{F}_n \subset \mathcal{F}$ are *mutually independent* if, for any $A_1 \in \mathcal{F}_1, \ldots, A_n \in \mathcal{F}_n$,

$$P(A_1 \cap A_2 \cap \ldots \cap A_n) = P(A_1)P(A_2) \ldots P(A_n).$$

Random variables $\xi_1, \ldots, \xi_n$ are said to be *mutually independent* if the σ-fields $\sigma(\xi_1), \ldots, \sigma(\xi_n)$ are mutually independent. Sometimes it also is convenient to say that a random variable ξ is independent of a σ-field $\mathcal{G}$ whenever the σ-fields $\sigma(\xi)$ and $\mathcal{G}$ are independent.

We denote the *expectation* of a random variable $\xi : \Omega \to \mathbf{R}^d$ by

$$E\xi = \int_\Omega \xi \, dP,$$

provided the above integral exists. We say that ξ is *integrable* if $E|\xi| < \infty$ and *square-integrable* if $E|\xi|^2 < \infty$, where $|\cdot|$ is the standard Euclidean norm in $\mathbf{R}^d$.

Let $\mathcal{G}$ be a σ-field on Ω such that $\mathcal{G} \subset \mathcal{F}$ and let $\xi : \Omega \to \mathbf{R}^d$ be an integrable random variable. A $\mathcal{G}$-measurable random variable with values in $\mathbf{R}^d$ is called the *conditional expectation of* ξ *given* $\mathcal{G}$ and denoted by $E(\xi|\mathcal{G})$ if

$$\int_A \xi \, dP = \int_A E(\xi|\mathcal{G}) \, dP \qquad \text{for all} \quad A \in \mathcal{G}.$$

The conditional expectation is unique to within equality P-a.s. and exists by virtue of the Radon-Nikodym theorem. It has the following properties:

(E1) $\xi \mapsto E(\xi|\mathcal{G})$ is a linear map,

(E2) $E(E(\xi|\mathcal{G})) = E(\xi)$,

(E3) $E(\xi \cdot \eta|\mathcal{G}) = \xi \cdot E(\eta|\mathcal{G})$ if ξ is $\mathcal{G}$-measurable (where $\cdot$ denotes the scalar product in $\mathbf{R}^d$),

(E4) $E(\xi|\mathcal{G}) = E(\xi)$ if ξ is independent of $\mathcal{G}$,

(E5) $E(E(\xi|\mathcal{G})|\mathcal{H}) = E(\xi|\mathcal{H})$ if $\mathcal{H}$ is a σ-algebra on Ω contained in $\mathcal{G}$.

A *stochastic process* with values in a topological space M is by definition a function $\xi : \Omega \times [0,\infty) \to M$ such that $\xi(\cdot, t)$ is a random variable for any $t \geq 0$. We shall often write $\xi(t)$ instead of $\xi(\cdot, t)$. We can also consider stochastic processes for which the interval $[0,\infty)$ is replaced by some other subset of $\mathbb{R}$, for example, a bounded interval or the set $\mathbb{N}$ of natural numbers. For any $\omega \in \Omega$, the function $[0,\infty) \ni t \mapsto \xi(\omega, t) \in M$ is called a *sample path* of ξ.

By a filtration $(\mathcal{F}_t)_{t\geq 0}$ we mean a family of σ-fields, all of which are contained in $\mathcal{F}$, such that $\mathcal{F}_s \subset \mathcal{F}_t$ if $s \leq t$. In addition, we will always assume that all the sets $A \in \mathcal{F}$ with $P(A) = 0$ belong to $\mathcal{F}_t$ for every $t \geq 0$ and the filtration is *right-continuous*, which means that $\mathcal{F}_t = \bigcap_{s>t} \mathcal{F}_s$ for every $t \geq 0$.

We say that a stochastic process $\xi : \Omega \times [0,\infty) \to M$ is *measurable* if the set $\{(\omega, t) : \xi(\omega, t) \in B\}$ belongs to the product σ-field $\mathcal{F} \otimes \mathcal{B})$ for any $B \in \mathcal{B}$. We say that a stochastic process ξ is *adapted* to the filtration $(\mathcal{F}_t)_{t\geq 0}$ if $\xi(t)$ is $\mathcal{F}_t$-measurable for every $t \geq 0$. Two stochastic processes ξ and η are called *stochastically equivalent* if $\xi(t) = \eta(t)$ P-a.s. for all $t \in [0,\infty)$, in which case the processes are also called *modifications* or *versions* of each other.

We say that a stochastic process $\xi : \Omega \times [0,\infty) \to M$ has the *Markov property* if

$$E\left(f(\xi(t))\middle|\sigma(\xi(u), u \leq s)\right) = E\left(f(\xi(t))\middle|\sigma(\xi(s))\right) \quad P\text{-a.s.} \qquad (2.1)$$

for any $0 \leq s \leq t$ and any bounded Borel-measurable function $f : M \to \mathbb{R}$.

Let $\xi : \Omega \times [0,\infty) \to \mathbb{R}^d$ be a stochastic process adapted to the filtration $(\mathcal{F}_t)_{t\geq 0}$. It is called a *martingale* (to be precise, an $(\mathcal{F}_t)$-*martingale*) if $\xi(t)$ is integrable for any $t \geq 0$ and

$$E\left(\xi(t)\middle|\mathcal{F}_s\right) = \xi(s) \quad P\text{-a.s.}$$

for any $0 \leq s \leq t$.

A *stopping time* is a random variable $\tau : \Omega \to [0,\infty]$ such that $\{\tau \leq t\} \in \mathcal{F}_t$ for every $t \in [0,\infty]$. For a stopping time τ, we put

$$\mathcal{F}_\tau = \{A \in \mathcal{F} : A \cap \{\tau \leq t\} \in \mathcal{F}_t \text{ for every } t \in [0,\infty]\}.$$

It is easily seen that $\mathcal{F}_\tau \subset \mathcal{F}$ is a σ-field on Ω. If $\xi : \Omega \times [0,\infty) \to M$ is a stochastic process, then $\xi^\tau : (\omega, t) \mapsto \xi(\omega, t \wedge \tau(\omega))$ is said to be the process *stopped* by τ. Here $\wedge$ denotes the minimum of two real numbers.

Definition 2.1. A stochastic process $w : \Omega \times [0, \infty) \to \mathbb{R}^d$ adapted to the filtration $(\mathcal{F}_t)_{t \geq 0}$ is called a *(standard) Brownian motion* (or *Wiener process*) in $\mathbb{R}^d$ if the sample paths $t \mapsto w(t)$ are P-a.s. continuous, $w(0) = 0$ P-a.s., and the increment $w(t) - w(s)$ is independent of $\mathcal{F}_s$ for any $0 \leq s \leq t$ and has normal distribution such that

$$P\{w(t) - w(s) \in A\} = \frac{1}{\left(\sqrt{2\pi(t-s)}\right)^d} \int_A \exp\left(-\frac{|x|^2}{2(t-s)}\right) dx \qquad (2.2)$$

for any Borel set $A \subset \mathbb{R}^d$, $|\cdot|$ being the standard Euclidean norm in $\mathbb{R}^d$.

Since $w(t) - w(s)$ is independent of $\mathcal{F}_s$, it is independent of $\mathcal{F}_r$ for any $0 \leq r \leq s$. It follows that w satisfies (2.1), that is, has the Markov property. Definition 2.1 also implies that the components $w^1, \ldots, w^d$ of w are mutually independent.

Another important consequence is that

$$E\left(w(t) \middle| \mathcal{F}_s\right) = w(s) \qquad (2.3)$$

$$E\left((w^i(t) - w^i(s))(w^j(t) - w^j(s)) \middle| \mathcal{F}_s\right) = t - s \qquad (2.4)$$

for any $0 \leq s \leq t$. An alternative way of expressing (2.3) and (2.4) is to say that $w^i(t)$ and $w^i(t)w^j(t) - t\delta^{ij}$ are martingales for any $i, j = 1, \ldots, d$. By a theorem of Lévy (see, for example, [Doo53], Chap. 7 or [KS88], Chap. 3), conditions (2.3) and (2.4) characterize Brownian motion.

More information on the notions mentioned in this section can be found, for example, in [Doo53], [IM64], [McK69], [GS72], [Fri75], [Arn74], [Hid80], [IW89], or [KS88]. These books, and a number of other books on stochastic processes and stochastic calculus, can also be consulted for the proofs of the existence of a Brownian motion in $\mathbb{R}^d$ and its properties.

2.2 Itô's Stochastic Integral

Integrals of the form $\int_0^t \xi(s)\, dw(s)$, where $\xi(t)$ is a stochastic process and $w(t)$ a Brownian motion, cannot be defined in the standard way as Lebesgue-Stieltjes integrals along each sample path $t \mapsto w(t)$, since the paths of a Brownian motion are well known to be P-a.s. nowhere differentiable and to have infinite variation on any finite interval $[0, t]$. The difficulty has been resolved by Itô [Itô87], [Itô44], who gave an elegant definition of what is now known as Itô's stochastic integral.

Here we shall present the definition of the stochastic integral withe respect to a Brownian motion $w(t)$ in $\mathbf{R}^d$. The integrand $\xi(t)$ will be a stochastic process of a certain class specified below with values in the set $(\mathbf{R}^d)^*$ of linear transformations from $\mathbf{R}^d$ to $\mathbf{R}$. We shall write $xy = x_i y^i$ for any $x = (x_i)$ in $(\mathbf{R}^d)^*$ and $y = (y^i)$ in $\mathbf{R}^d$. By $|\cdot|$ we shall designate either the absolute value of a real number or the standard Euclidean norm in $(\mathbf{R}^d)^*$, depending on the context. In the latter case $|x|^2 = \delta^{ij} x_i x_j$ for any $x = (x_i)$ in $(\mathbf{R}^d)^*$.

Definition 2.2. The following spaces of processes are involved in the construction of Itô's stochastic integral:

a) The space $\mathcal{L}_0$ of simple processes $\xi : \Omega \times [0, \infty) \to (\mathbf{R}^d)^*$. A *simple process* is by definition a stochastic process adapted to the filtration $(\mathcal{F}_t)_{t \geq 0}$, bounded P-a.s., and such that there exists an increasing sequence of real numbers $0 = t_0 < t_1 < \ldots < t_n < \ldots$ with $t_n \to \infty$ as $n \to \infty$ such that

$$\xi(t) = \xi(t_i) \quad \text{for any} \quad t \in (t_{i-1}, t_i], \tag{2.5}$$

where $i = 1, 2, \ldots$.

b) The space $\mathcal{L}^2$ of all measurable stochastic processes $\xi : \Omega \times [0, \infty) \to (\mathbf{R}^d)^*$ adapted to the filtration $(\mathcal{F}_t)_{t \geq 0}$ such that for every $t \geq 0$,

$$\|\xi\|^2_{\mathcal{L}^2,t} = E \int_0^t |\xi(s)|^2 \, ds < \infty.$$

c) The space $\mathcal{M}^{2,c}$ of *continuous square-integrable martingales* $\xi : \Omega \times [0, \infty) \to \mathbf{R}$, that is, martingales with P-a.s. continuous sample paths such that for every $t \geq 0$,

$$\|\xi\|^2_{\mathcal{M}^{2,c},t} = E|\xi(t)|^2 < \infty.$$

d) The space $\mathcal{P}^2$ of all measurable stochastic processes $\xi : \Omega \times [0, \infty) \to (\mathbf{R}^d)^*$ adapted to the filtration $(\mathcal{F}_t)_{t \geq 0}$ such that for every $t \geq 0$,

$$\int_0^t |\xi(s)|^2 \, ds < \infty \quad P\text{-a.s.}$$

e) The space $\mathcal{M}^{2,c}_{\mathrm{loc}}$ of *continuous local martingales* $\xi : \Omega \times [0,\infty) \to \mathbb{R}$, which are defined as processes with P-a.s. continuous sample paths such that there exists a sequence τ_n of stopping times such that $P\{\tau_n \leq t\} \to 0$ as $n \to \infty$ for all $t \geq 0$, $\tau_n \leq \tau_{n+1}$, and the stopped process $\xi^{\tau_n}(t) = \xi(t \wedge \tau_n)$ is a martingale for each n.

The spaces $\mathcal{L}^2$ and $\mathcal{M}^{2,c}$ are equipped with the metrics

$$\rho_{\mathcal{L}^2}(\xi,\eta) = \sum_{n=1}^{\infty} 2^{-n}\left(1 \wedge \|\xi - \eta\|_{\mathcal{L}^2,n}\right), \tag{2.6}$$

$$\rho_{\mathcal{M}^{2,c}}(\xi,\eta) = \sum_{n=1}^{\infty} 2^{-n}\left(1 \wedge \|\xi - \eta\|_{\mathcal{M}^{2,c},n}\right). \tag{2.7}$$

In $\mathcal{P}^2$ and $\mathcal{M}^{2,c}_{\mathrm{loc}}$ we define the metrics

$$\rho_{\mathcal{P}^2}(\xi,\eta) = \sum_{n=1}^{\infty} 2^{-n} E\left(1 \wedge \left[\int_0^t |\xi(s) - \eta(s)|^2\, ds\right]^{1/2}\right), \tag{2.8}$$

$$\rho_{\mathcal{M}^{2,c}_{\mathrm{loc}}}(\xi,\eta) = \sum_{n=1}^{\infty} 2^{-n} E\left(1 \wedge \sup_{s\in[0,t]} |\xi(s) - \eta(s)|\right). \tag{2.9}$$

Remark 2.1. The choice of simple processes as ones with left-continuous sample paths is not essential for the construction of Itô's integral with respect to a Brownian motion. Right-continuous sample paths are also admissible.

Remark 2.2. To be precise, processes $\xi, \eta \in \mathcal{L}^2$ should be identified if $\rho_{\mathcal{L}^2}(\xi,\eta) = 0$, and $\mathcal{L}^2$ should be regarded as a space of equivalence classes. Similar remarks apply to $\mathcal{P}^2$, $\mathcal{M}^{2,c}$, and $\mathcal{M}^{2,c}_{\mathrm{loc}}$.

Remark 2.3. A sequence $\xi_n \in \mathcal{P}^2$ converges to $\xi \in \mathcal{P}^2$ in the metric $\rho_{\mathcal{P}^2}$ if and only if $\int_0^t |\xi_n(s) - \xi(s)|^2\, ds \to 0$ in probability for every $t \geq 0$.

Remark 2.4. A sequence $\xi_n \in \mathcal{M}^{2,c}_{\mathrm{loc}}$ converges to $\xi \in \mathcal{M}^{2,c}_{\mathrm{loc}}$ in the metric $\rho^{\mathcal{M}^{2,c}_{\mathrm{loc}}}$ if and only if $\sup_{s\in[0,t]} |\xi_n(s) - \xi(s)| \to 0$ in probability for every $t \geq 0$, in which case the sequence is said to converge in probability, uniformly on compact sets.

Theorem 2.1. *a) $\mathcal{L}_0$ is a dense subspace of the metric space $\mathcal{L}^2$ with metric $\rho_{\mathcal{L}^2}$;*

b) The metric space $\mathcal{M}^{2,c}$ with metric $\rho_{\mathcal{M}^{2,c}}$ is complete.

c) $\mathcal{L}^2$ is a dense subspace of the metric space $\mathcal{P}^2$ with metric $\rho_{\mathcal{P}^2}$.

d) $\mathcal{M}^{2,c}_{\text{loc}}$ with metric $\rho_{\mathcal{M}^{2,c}_{\text{loc}}}$ is a complete metric space.

These results are now standard in stochastic calculus. For the proofs of a) and b) we refer, for example, to [IW89] or [KS88]. The proof of c) can be found in [Fri75] or [LS77]. Assertion d) follows from Theorem 2.1.2 of [Kun90].

We are now ready to construct the stochastic integral, first as a functional $I : \mathcal{L}^2 \to \mathcal{M}^{2,c}$, and then to extend it to a functional $I : \mathcal{P}^2 \to \mathcal{M}^{2,c}_{\text{loc}}$. For a simple process $\xi \in \mathcal{L}_0$ defined by (2.5) we put

$$(I\xi)(t) = \sum_{i=0}^{\infty} \xi(t_i)\left[w(t \wedge t_{i+1}) - w(t \wedge t_i)\right] \tag{2.10}$$

for any $t \geq 0$. Note that there are only finitely many non-zero terms in the sum for every $t \geq 0$. It is easily verified using (2.3) and (2.4) that for any $\xi, \eta \in \mathcal{L}_0$, and $0 \leq s \leq t$,

$$E\left((I\xi)(t)|\mathcal{F}_s\right) = (I\xi)(s) \tag{2.11}$$

$$E\left(|(I\xi)(t) - (I\xi)(s)|^2|\mathcal{F}_s\right) = E\left(\int_s^t |\xi(r) - \xi(s)|^2 dr|\mathcal{F}_s\right) \tag{2.12}$$

It follows from (2.11) that $I\xi$ is a martingale for any $\xi \in \mathcal{L}_0$. It is seen from (2.10) that it has P-a.s. continuous sample paths, and from (2.12) that it is square-integrable. Thus $I : \mathcal{L}_0 \to \mathcal{M}^{2,c}$. Formula (2.10) implies that I is a linear mapping. By taking the expectation of both sides of (2.12), we find that $\|I\xi\|_{\mathcal{M}^{2,c},t} = \|\xi\|_{\mathcal{L}^2,t}$ for any $t \geq 0$, which yields $\rho_{\mathcal{L}^2}(\xi,\eta) = \rho_{\mathcal{M}_{2,c}}(\xi,\eta)$ for any $\xi, \eta \in \mathcal{L}_0$. This means that I is a linear isometry from the dense subspace $\mathcal{L}_0$ of $\mathcal{L}_2$ to the complete space $\mathcal{M}^{2,c}$. Therefore it has a unique extension to a linear isometry from $\mathcal{L}^2$ to $\mathcal{M}^{2,c}$, which will be denoted by the same symbol I.

Definition 2.3. The linear isometry $I : \mathcal{L}^2 \to \mathcal{M}^{2,c}$ is called the *Itô stochastic integral on $\mathcal{L}^2$*. We shall often write $\int_0^t \xi(s)\, dw(s)$ or just $\int_0^t \xi\, dw$ instead of $(I\xi)(t)$.

A simple argument shows that formulae (2.11) and (2.12) are true for any $\xi \in \mathcal{L}^2$.

We need the following lemma to define the stochastic integral as a functional $I : \mathcal{P}^2 \to \mathcal{M}^{2,c}_{\text{loc}}$ by extending it from $\mathcal{L}^2$.

Lemma 2.2. *The functional* $I : \mathcal{P}^2 \supset \mathcal{L}^2 \to \mathcal{M}_2 \subset \mathcal{M}_{\text{loc}}^{2,c}$ *is uniformly continuous with respect to the metrics* $\rho_{\mathcal{P}^2}$ *and* $\rho_{\mathcal{M}_{\text{loc}}^{2,c}}$ *induced on* $\mathcal{L}^2$ *and* $\mathcal{M}_{2,c}$ *from* $\mathcal{P}^2$ *and* $\mathcal{M}_{\text{loc}}^{2,c}$*, respectively.*

Proof. The lemma is a simple consequence of the inequality

$$P\left\{\sup_{s\in[0,t]} |(I\xi)(s)| > \varepsilon\right\} \leq \frac{\mu}{\varepsilon^2} + P\left\{\int_0^t |\xi(s)|^2\, ds > \mu\right\}$$

valid for every $\xi \in \mathcal{L}^2$ and $\varepsilon, \mu > 0$. The inequality is proved, for example, in [LS77], Chap. 4 and [Fri75], Chap. 4. ∎

As a uniformly continuous functional from $\mathcal{L}^2$, which is a dense subspace of $\mathcal{P}^2$ in the $\rho_{\mathcal{P}^2}$ metric, to the complete metric space $\mathcal{M}_{\text{loc}}^{2,c}$, I has a unique extension to a continuous functional from $\mathcal{P}^2$ to $\mathcal{M}_{\text{loc}}^{2,c}$, which we shall denote by the same symbol I. It is readily verified that this extended functional is also linear.

Definition 2.4. The extended linear functional $I : \mathcal{P}^2 \to \mathcal{M}_{\text{loc}}^{2,c}$ is called the *stochastic integral on* $\mathcal{P}^2$. We shall often write $\int_0^t \xi(s)\, dw(s)$ or $\int_0^t \xi\, dw$ instead of $(I\xi)(t)$ for any $\xi \in \mathcal{P}^2$.

Note that when $\xi \in \mathcal{P}^2$, the integral $I\xi$ will, in general, no longer satisfy (2.11) or (2.12).

2.3 Itô Processes. Itô Formula

In stochastic calculus we frequently encounter processes $\xi : \Omega \times [0, \infty) \to \mathbb{R}$ of the form

$$\xi(t) - \xi(0) = \int_0^t a(s)\, dw(s) + \int_0^t b(s)\, ds, \tag{2.13}$$

where $a : \Omega \times [0, \infty) \to (\mathbb{R}^d)^*$ and $b : \Omega \times [0, \infty) \to \mathbb{R}$ are measurable stochastic processes adapted to the filtration $(\mathcal{F}_t)_{t\geq 0}$ such that

$$\int_0^t |a(s)|^2\, ds \quad < \quad \infty \qquad P\text{-a.s.} \tag{2.14}$$

$$\int_0^t |b(s)|\, ds \quad < \quad \infty \qquad P\text{-a.s.} \tag{2.15}$$

for all $t \geq 0$. This means that a is a process of class $\mathcal{P}^2$ defined in Sect. 2.2. Similarly, we shall denote by $\mathcal{P}_1$ the class of all processes that satisfy the above conditions for b.

Definition 2.5. We call a process $\xi : \Omega \times [0, \infty) \to \mathbf{R}$ of the form (2.13), where $a \in \mathcal{P}^2$ and $b \in \mathcal{P}^1$, an *Itô process.*

Note that every Itô process is measurable, adapted to the filtration $(\mathcal{F}_t)_{t \geq 0}$, and has P-a.s. continuous sample paths. For example, each of the components w^i of a Brownian motion in $\mathbf{R}^d$ is an Itô process, and so is any deterministic (that is, independent of $\omega \in \Omega$) real-valued process with absolutely continuous paths. A multitude of other examples can be obtained with the aid of the Itô formula below (Theorem 2.3).

It proves convenient to define the following stochastic integrals withe respect to Itô processes.

Definition 2.6. Let ξ^i, where $i = 1, 2, \ldots$ be Itô processes such that

$$\xi^i(t) - \xi^i(0) = \int_0^t a^i(s)\, dw(s) + \int_0^t b^i(s)\, ds$$

with $a^i \in \mathcal{P}^2$ and $b^i \in \mathcal{P}^1$, and let η be a measurable process adapted to the filtration $(\mathcal{F}_t)_{t \geq 0}$ with P-a.s. continuous sample paths. We put

$$\int_0^t \eta\, d\xi^i = \int_0^t \eta(s) a^i(s)\, dw(s) + \int_0^t \eta(s) b^i(s)\, ds, \quad (2.16)$$

$$\int_0^t \eta\, d\xi^i d\xi^j = \int_0^t \eta(s) \langle a^i(s), a^j(s) \rangle\, ds, \quad (2.17)$$

$$\int_0^t \eta\, d\xi^{i_1} \ldots d\xi^{i_n} = 0, \quad n \geq 3 \quad (2.18)$$

for any $t \geq 0$. Here $\langle x, y \rangle = \delta^{ij} x_i y_j$ is the ordinary Euclidean scalar product of $x = (x_i)$ and $y = (y_i)$ in $(\mathbf{R}^d)^*$.

The condition that η should have P-a.s. continuous paths can be relaxed. For the integrals on the right-hand sides of (2.16) and (2.17) to exist it suffices that $\eta a^1 \in \mathcal{P}^2$, $\eta b^1 \in \mathcal{P}^1$, and $\eta \langle a^1, a^2 \rangle \in \mathcal{P}_1$.

The following *Itô rules* follow directly from (2.16)-(2.18):

$$\text{a)} \int_0^t dw^i dw^j = \delta^{ij} t, \quad \text{b)} \int_0^t dw^i dt = 0, \quad \text{c)} \int_0^t dt\, dt = 0. \quad (2.19)$$

Here w^i are the components of a Brownian motion in $\mathbf{R}^d$.

Theorem 2.3 (Itô formula). *Let* $\xi^1, \ldots, \xi^n : \Omega \times [0, \infty) \to \mathbf{R}$ *be Itô processes and let* $f : \mathbf{R}^n \to \mathbf{R}$ *be a function of class* C^2. *We put* $\xi = (\xi^1, \ldots, \xi^n)$. *Then* $f(\xi)$ *is an Itô process and*

$$f(\xi(t)) - f(\xi(0)) = \int_0^t \partial_i f(\xi)\, d\xi^i + \frac{1}{2} \int_0^t \partial_i \partial_j f(\xi)\, d\xi^i d\xi^j \tag{2.20}$$

for any $t \geq 0$. *The integrals on the right-hand side of* (2.20) *exist, since* $\partial_i f(\xi)$ *and* $\partial_i \partial_j f(\xi)$ *have* P*-a.s. continuous sample paths,* $\partial_i f$ *and* $\partial_i \partial_j f$ *being the partial derivatives of* f.

The proof of Itô's formula can be found, for example, in [McK69], [GS72], [Fri75], [IW89], [KS88], and almost any other book on stochastic calculus.

It is customary to suppress the integral sign in equalities involving linear combinations of stochastic integrals on both sides. For example, since $f(\xi(t)) - f(\xi(0)) = \int_0^t df(\xi)$, the Itô formula (2.20) can be written as

$$df(\xi) = \partial_i f(\xi)\, d\xi^i + \frac{1}{2} \partial_i \partial_j f(\xi)\, d\xi^i d\xi^j. \tag{2.21}$$

This convention, referred to as the *stochastic differential notation*, will often be used in the sequel. In this notation the Itô rules (2.19) will be written as

$$\text{a) } dw^i dw^j = \delta^{ij} dt, \quad \text{b) } dw^i dt = 0, \quad \text{c) } dt\, dt = 0.$$

2.4 Stratonovich Integrals

The Itô formula (2.20) or (2.21) can be regarded as a stochastic analogue of the change-of-variables rule in ordinary calculus. The second term on the right-hand side of (2.20) or (2.21) is a purely stochastic feature known as the *stochastic correction*. Replacing the Itô integral by the Stratonovich integral defined below, one can dispense with the correction term (more precisely, absorb it into the integral). This is especially important when developing stochastic calculus on manifolds to ensure the correct transformation properties. Another property of Stratonovich integrals, which will also prove crucial to us, is their robust behaviour under approximation by processes with piecewise smooth paths.

Definition 2.7. Let ξ and ζ be Itô processes. The *Stratonovich integral* is defined by

$$\int_0^t \xi \circ d\zeta = \int_0^t \xi\, d\zeta + \frac{1}{2} \int_0^t d\xi\, d\zeta,$$

or, in the stochastic differential notation,

$$\xi \circ d\zeta = \xi\, d\zeta + \frac{1}{2} d\xi\, d\zeta.$$

The corollary below, which follows immediately from Theorem 2.3, states the Itô formula in terms of the Stratonovich integral.

Corollary 2.4. *If $\xi = (\xi^1, \dots, \xi^n)$, where $\xi^1, \dots, \xi^n$ are Itô processes, and $f : \mathbf{R}^n \to \mathbf{R}$ is of class C^3, then*

$$f(\xi(t)) - f(\xi(0)) = \int_0^t \partial_i f(\xi) \circ d\xi^i, \tag{2.22}$$

or, equivalently, $df(\xi) = \partial_i f(\xi) \circ d\xi^i$.

Note that more smoothness is required of f in (2.22) than in (2.20). This is necessary to ensure that $\partial_i f(\xi)$ is an Itô process.

2.5 Stochastic Differential Equations on Manifolds

We begin this section with the definition of a strong solution of a stochastic differential equation (SDE) in $\mathbf{R}^n$. We confine ourselves to time-homogeneous equations. Then we proceed to discussing SDEs on manifolds, in which case the notion of a solution will be extended so as to include solutions that admit explosions.

Definition 2.8. Let $w = (w^1, \dots, w^d)$ be a Brownian motion in $\mathbf{R}^d$ and let $a^i : \mathbf{R}^n \to (\mathbf{R}^d)^*$ and $b^i : \mathbf{R}^n \to \mathbf{R}$ be Borel-measurable functions for $i = 1, \dots, n$. We say that $\xi = (\xi^1, \dots, \xi^n) : \Omega \times [0, \infty) \to \mathbf{R}^n$, where $\xi^1, \dots, \xi^n$ are Itô processes, is a *strong solution* of the SDE

$$d\xi^i(t) = a^i(\xi(t))\, dw(t) + b^i(\xi(t))\, dt \tag{2.23}$$

if $a^i(\xi(\cdot)) \in \mathcal{P}^2$, $b^i(\xi(\cdot)) \in \mathcal{P}^1$, and

$$\xi^i(t) - \xi^i(0) = \int_0^t a^i(\xi(s))\, dw(s) + \int_0^t b^i(\xi(s))\, ds \quad P\text{-a.s.}$$

for any $i = 1, \dots, n$ and $t \geq 0$.

Theorem 2.5. *Suppose that a^i and b^i in Definition 2.8 satisfy the Lipschitz and bounded growth conditions*

$$|a^i(x) - a^i(y)| \leq K|x-y|, \quad |a^i(x)| \leq K(1+|x|), \tag{2.24}$$
$$|b^i(x) - b^i(y)| \leq K|x-y|, \quad |b^i(x)| \leq K(1+|x|), \tag{2.25}$$

for every $x, y \in \mathbf{R}^n$ and some $K > 0$, and let $\eta : \Omega \to \mathbf{R}^n$ be an $\mathcal{F}_0$-measurable random variable such that $E|\eta|^2 < \infty$. Then there exists a unique (to within equality P-a.s.) strong solution ξ of the SDE (2.23) *with initial condition $\xi(0) = \eta$. Moreover, $E|\xi(t)|^2 < \infty$ for every $t \geq 0$.*

The proof of this standard result can be found, for example, in [LS77], [Fri75], [IW89], or [KS88].

Example 2.1. Let w be a one-dimensional Brownian motion and let $a, b \in \mathbf{R}$ be constants. It is easily verified that the unique solution to the SDE

$$d\xi(t) = a\,dw(t) - b\xi(t)\,dt$$

with initial condition $\xi(0) = x \in \mathbf{R}$ is the *Ornstein-Uhlenbeck velocity process*

$$\xi(t) = e^{-bt}x + ae^{-bt}\int_0^t e^{bs}\,dw(s).$$

Example 2.2. The unique solution to the SDE $d\xi(t) = \xi(t)\,dw(t)$ with initial condition $\xi(0) = 1$, w being a one-dimensional Brownian motion, is the so-called *exponential martingale*

$$\xi(t) = \exp\left[w(t) - t/2\right].$$

According to Definition 2.8, a strong solution to a SDE must be defined for all $t \geq 0$. This condition proves too stringent for SDEs on manifolds, in which case one frequently encounters solutions that 'explode' (that is, escape to infinity in finite time) with non-zero probability. Note that the standard global Lipschitz and bounded growth conditions (2.24) and (2.25), which ensure the existence and uniqueness of a solution for all $t \geq 0$, do not readily generalize to manifolds. As solutions of a SDE on a manifold M we must therefore admit stochastic processes defined on a random interval $[0, \tau)$, where τ is a stopping time.

Definition 2.9. Let $w = (w^1, \ldots, w^d)$ be a Brownian motion in $\mathbb{R}^d$ and let $A_0, A_1, \ldots, A_d \in \mathcal{X}$ be vector fields on a finite-dimensional manifold M. A process $\xi : \Omega \times [0, \tau) \to M$ with $\tau : \Omega \to [0, \infty]$ a stopping time such that

a) $0 < \tau$ P-a.s.,

b) $\xi(t) : \Omega \supset \{t < \tau\} \to M$ is $\mathcal{F}_t$-measurable for every $t \geq 0$,

c) the sample paths $[0, \tau) \ni t \mapsto \xi(t) \in M$ are P-a.s. continuous,

is called a *solution* of the SDE

$$d\xi(t) = A_0(\xi(t))\,dt + A_i(\xi(t)) \circ dw^i(t) \tag{2.26}$$

if for every $f \in C_0^\infty(M)$ and $t \geq 0$,

$$f(\xi(t)) - f(\xi(0)) = \int_0^t (A_0 f)(\xi(s))\,ds + \int_0^t (A_i f)(\xi(s)) \circ dw^i(s) \tag{2.27}$$

P-a.s. on $\{t < \tau\}$, $C_0^\infty(M)$ being the set of smooth functions $f : M \to \mathbb{R}$ with compact support.

Remark 2.5. We shall often refer to the SDE (2.26) by writing it in local coordinates as

$$d\xi^j(t) = A_0^j(\xi(t))\,dt + A_i^j(\xi(t)) \circ dw^i(t),$$

where A_i^j are the components of the vector fields A_i.

The standard theorem on the existence and uniqueness of solutions of SDEs on manifolds reads

Theorem 2.6. *For any $\mathcal{F}_0$-measurable random variable $\eta : \Omega \to M$ there exists a solution $\xi : \Omega \times [0, \tau) \to M$ to the SDE* (2.26) *with initial condition $\xi(0) = \eta$ such that if $\tilde{\xi} : \Omega \times [0, \tilde{\tau}) \to M$ is another solution to* (2.26) *with $\xi(0) = \tilde{\xi}(0)$ P-a.s., then*

$$\tau \geq \tilde{\tau} \quad \text{and} \quad \xi = \tilde{\xi} \quad \text{on } [0, \tilde{\tau}) \quad P\text{-a.s.} \tag{2.28}$$

The proof of Theorem 2.6 can be extracted from [IW89], [Elw82], or [Kun90].

Definition 2.10. A solution $\xi : \Omega \times [0, \tau) \to M$ to (2.26) that satisfies condition (2.28) is called a *maximal solution* and the corresponding stopping time τ is called the *explosion time* of ξ.

The theorem below is concerned with the behaviour of a maximal solution to a SDE near explosion time. It is a generalization of the well-known result for ordinary differential equations and, in fact, justifies the term 'explosion time.' Given a function $f : [a,b) \to M$, we shall write $\lim_{t \nearrow b} f(t) = \infty$ if for any compact set $K \subset M$ there is $s \in [a,b)$ such that $f(t) \in M \setminus K$ for all $t \in [s,b)$.

Theorem 2.7. *If $\xi : \Omega \times [0,\tau) \to M$ is a maximal solution to (2.26), then*

$$P\left\{\tau = \infty \quad \text{or} \quad \lim_{t \nearrow \tau} \xi(t) = \infty\right\} = 1.$$

The proof of the above result follows, for example, from Theorem VII.6 of [Elw82].

An important consequence of Theorem 2.7 is that for any maximal solution $\xi : \Omega \times [0,\tau) \to M$ of (2.26) on a compact manifold M, $\tau = \infty$ P-a.s., that is, there are no explosions.

The above definitions and theorems can readily be extended to the case of initial conditions at $s \geq 0$. Given any $s \geq 0$ and $x \in M$, we denote by $t \mapsto \xi(s,x;t)$ the maximal solution to the SDE (2.26) with initial condition $\xi(s,x;s) = x$. Let $\tau(s,x)$ be the corresponding explosion time.

Theorem 2.8. *The system of maximal solutions $\xi(s,x;\cdot) : \Omega \times [s,\tau(s,x)) \to M$, where $s \geq 0$ and $x \in M$, has a version (for which we retain the same symbols $\xi(s,x;t)$ and $\tau(s,x)$) such that*

a) $\xi(s,\xi(r,x;s);t) = \xi(r,x;t)$ P-a.s. on $\{t < \tau(r,x)\}$ for any $0 \leq r \leq s \leq t$;

b) $M(s,t) = \{x \in M : t < \tau(s,x)\}$ is an open subset of M for every $0 \leq s < t$ and $\xi(s,\cdot\,;t) : M(s,t) \to M$ is a C^∞-diffeomorphism onto an open subset of M, P-a.s.;

c) $\xi(s,x;t)$ is P-a.s. continuous as a function of (s,x,t), and so are its differentials of any order withe respect to x.

For the proof we refer the reader to Sect. VIII.2 of [Elw82] and Sect. 4.8 of [Kun90].

From now on we shall always use the version of $\xi(s,x;t)$ described in Theorem 2.8 and refer to it as the *stochastic flow* on M associated to the SDE (2.26).

An important consequence of Theorem 2.8 a) is that for any fixed $r \geq 0$ and $x \in M$, $\xi(r, x; \cdot)$ satisfies the Markov property (2.1) for any s and t such that $r \leq s \leq t$, P-a.s. on $\{t < \tau(r, x)\}$.

Finally, we introduce the notion of the differential generator of $\xi(s, x; t)$. On evaluating the Stratonovich integral in (2.27), we find that for any $f \in C_0^\infty(M)$

$$\begin{aligned} f(\xi(s,x;t)) - x &= \int_s^t \left(\frac{1}{2}\delta^{ij} A_i A_j f + A_0 f\right)(\xi(s,x;u))\, du \quad (2.29) \\ &\quad + \int_s^t (A_i f)(\xi(s,x;u))\, dw(u) \end{aligned}$$

P-a.s. on $\{t < \tau(s, x)\}$. We extend $f(\xi(s, x; t))$ to a process defined for all $t \geq s$ by setting it equal to zero if $t \geq \tau(s, x)$. By Theorem 2.7, since f has compact support, this yields a process with P-a.s. continuous trajectories. The processes $(A_i f)(\xi(s, x; t))$ and $(\frac{1}{2}\delta^{ij} A_i A_j f + A_0 f)(\xi(s, x; t))$ can be extended to any $t \geq s$ in the same way. We retain the same notation for the extended processes. Then (2.29) becomes an equality satisfied P-a.s. on the whole set Ω. By taking the expectation of both sides of (2.29), we get

$$Ef(\xi(s,x;t)) - x = E\int_s^t \left(\frac{1}{2}\delta^{ij} A_i A_j f + A_0 f\right)(\xi(s,x;u))\, du.$$

It follows that

$$(Df)(x) = \lim_{t \searrow s} \frac{Ef(\xi(s,x;t)) - x}{t - s} = \left(\frac{1}{2}\delta^{ij} A_i A_j f + A_0 f\right)(x).$$

The operator $D = \frac{1}{2}\delta^{ij} A_i A_j + A_0$ is called the *differential generator* of $\xi(s, x; t)$.

Chapter 3

Stochastic Development on Finsler Spaces

3.1 Riemannian Stochastic Development

In this section M will be a d-dimensional Riemannian manifold with metric tensor $g(x), x \in M$. The coefficients of the corresponding Levi-Civita connection ∇ will be denoted by

$$\gamma^i_{jk} = \frac{1}{2} g^{in} \left(\partial_j g_{nk} + \partial_k g_{jn} - \partial_n g_{jk} \right). \tag{3.1}$$

3.1.1 Deterministic Case

Let x be an integral curve of a vector field X on M, that is,

$$\frac{dx}{dt} = X(x).$$

A vector field Y on M is said to be *parallel* along x if

$$\left(\nabla_X Y\right)(x) = 0. \tag{3.2}$$

Putting $y = Y(x)$, we can write (3.2) in local coordinates as

$$\frac{dy^i}{dt} + \gamma^i_{jk}(x) y^j \frac{dx^k}{dt} = 0. \tag{3.3}$$

The *horizontal lift* of a vector field $X = X^i \partial_i$ from M to TM is a vector field $h(X)$ on TM defined by

$$h(X)(x, y) = X^i(x) \left(\partial_i - \gamma^k_{ij}(x) y^j \dot{\partial}_k \right).$$

An equivalent way of stating condition (3.2) is to say that $\frac{dy}{dt}$ is the horizontal lift of X in the sense that,

$$\frac{dy}{dt} = h(X)(x, y). \tag{3.4}$$

For a smooth curve (x, y) on TM such that (3.4) holds we say that y is a *horizontal lift* of x to TM. Such a y is defined uniquely for each initial condition $y(0) \in TM_{x(0)}$. Equation (3.4) implies, in particular, that $y(t) \in TM_{x(t)}$ and

$$\frac{dy^i}{dt} = -\gamma^i_{jk}(x) y^j \frac{dx^k}{dt} \tag{3.5}$$

in local coordinates.

The mapping

$$\Pi_{s,t} : TM_{x(s)} \ni v \mapsto y(s, v; t) \in TM_{x(t)}$$

where $y(s, v; \cdot)$ is the horizontal lift of x to TM such that $y(s, v; s) = v$, is called the *h-parallel transport* along x. The following proposition means that $\Pi_{s,t}$ is a unitary mapping from $TM_{x(s)}$ to $TM_{x(t)}$.

Proposition 3.1. *Let x be a smooth curve on M. If y and $\tilde{y}$ are horizontal lifts of x to TM, then*

$$g_{ij}(x) y^i \tilde{y}^j = \text{const}\,.$$

Proof. Differentiate $g_{ij}(x) y^i \tilde{y}^j$ with respect to t, use (3.3) to express $\frac{dy^i}{dt}$ and $\frac{d\tilde{y}^j}{dt}$, and apply the metricity condition of the Levi-Civita connection

$$\nabla_k g_{ij} = \partial_k g_{ij} - g_{in} \gamma^n_{kj} - g_{nj} \gamma^n_{ik} = 0 \tag{3.6}$$

(which follows from (3.1)) to show that

$$\frac{d}{dt} \left(g_{ij}(x) y^i \tilde{y}^j \right) = 0,$$

which proves the claim. ∎

Let OM be the orthonormal frame bundle over M and X a vector field on M. The *horizontal lift* of X to OM is a vector field $l(X)$ on OM defined by

$$l(X)(x, z) = X^i(x) \left(\partial_i - \gamma^k_{ij}(x) z^j_n \partial^n_k \right).$$

Let x be a smooth curve on M and (x, z) a smooth curve on OM. We shall call z a *horizontal lift* of x to OM if

$$\frac{dz}{dt} = l(A)(x, z). \tag{3.7}$$

Here A is a vector field on M such that x is an integral curve of A. In local coordinates equation (3.7) takes the form

$$\frac{dz_n^i}{dt} = -\gamma_{jk}^i z_n^j \frac{dx^k}{dt}, \tag{3.8}$$

where $z_1, \ldots, z_d$ are the vectors forming the frame z. An equivalent condition is that each vector z_n is a horizontal lift of x to TM, that is,

$$\frac{dz_n}{dt} = h(A)(x, z_n) \tag{3.9}$$

for $n = 1, \ldots, d$. However, this equivalence will no longer be valid in the case of a Finsler manifold.

The lift from M to OM makes it possible to lift any smooth curve x from M to any tensor bundle $T_m^n M$. Namely, for a smooth curve (x, u) on $T_m^n M$ we say that u is a *horizontal lift* of x to $T_m^n M$ if for any horizontal lift z of x to OM the tensor $u(t)$ has constant components in the frame $z(t)$. This horizontal lift u depends on the initial condition $u(0) \in T_m^n M_{x(0)}$, but not on the choice of z. It is clear that for $TM = T_0^1 M$ we obtain the same horizontal lift of x as that defined by (3.4).

The *parallel transport* of tensors along x is defined by

$$\Pi_{s,t} : T_m^n M_{x(s)} \ni v \mapsto u(s, v; t) \in T_m^n M_{x(t)},$$

where $u(s, v; \cdot)$ is the horizontal lift of x to $T_m^n M$ such that $u(s, v; s) = v$.

Let M and $\tilde{M}$ be d-dimensional Riemannian manifolds. Suppose that they start rolling against one another without slipping, the point of contact tracing smooth curves x and $\tilde{x}$ on M and $\tilde{M}$, respectively. We take horizontal lifts z and $\tilde{z}$ of x and $\tilde{x}$ to OM and $O\tilde{M}$. The condition that no slipping is involved means that the components of the velocity vector $\frac{dx}{dt}$ in the frame z are the same as those of $\frac{d\tilde{x}}{dt}$ in $\tilde{z}$. This can be expressed by the differential equation

$$\zeta_j^i \frac{dx^j}{dt} = \tilde{\zeta}_j^i \frac{d\tilde{x}^j}{dt}, \tag{3.10}$$

where ζ and $\tilde{\zeta}$ are the dual frames to z and $\tilde{z}$. If the initial conditions $x(0), z(0)$ and $\tilde{x}(0), \tilde{z}(0)$ are fixed, then this defines a one-to-one mapping between smooth curves x and $\tilde{x}$ on the two manifolds, which is known as (*Riemannian*) *rolling.*

A typical situation is when $\tilde{M} = \mathbf{R}^d$. In this case $\tilde{z}$ is constant and can be chosen so that $\tilde{z}^i_n = \delta^i_n$ (the canonical basis). Then equations (3.10) and (3.8) reduce to

$$\frac{dx^i}{dt} = z^i_n \frac{d\tilde{x}^n}{dt} \tag{3.11a}$$

$$\frac{dz^i_n}{dt} = -\gamma^i_{jk} z^j_n z^k_m \frac{d\tilde{x}^m}{dt}. \tag{3.11b}$$

For any initial conditions $x(0), z(0)$ the system (3.11) defines a smooth curve (x, z) on OM called the (*Riemannian*) *development of* $\tilde{x}$. In particular, if $\tilde{x}$ is a straight line in $\mathbf{R}^d$, then x will be a geodesic on M.

3.1.2 Stochastic Case

Let M be a d-dimensional Riemannian manifold M. In the previous section we considered the horizontal lift of a smooth curve from M to TM. This will now be extended to the case when the curve is replaced by a diffusion x on M. We shall approximate the diffusion by a stochastic process with piecewise smooth sample paths, take the deterministic horizontal lift of each smooth piece of each path from M to TM, paste the pieces together to form continuous paths on TM, and consider the limit of the resulting process.

An *interpolation rule* is a measurable mapping $I : V \times [0,1] \to M$, where V is a neighbourhood of the diagonal $\{(x,x) : x \in M\}$ in $M \times M$, such that

(I1) $I(x,x,t) = x$ for all $t \in [0,1]$,

(I2) $I(x,x',0) = x$ and $I(x,x',t) = x'$,

(I3) $\partial^m I(x,x',t)/dt^m = O(d^m(x,x'))$ for $m = 1,2,3$,

where $d(x,x')$ is the distance between x and x' measured along the shortest geodesic joining the two points.

There is a multitude of interpolation rules to choose from. A natural one, which will be used in what follows, is the *geodesic interpolation rule* such that $t \mapsto I(x,x',t)$ is the shortest geodesic on M starting from x at $t = 0$ and arriving at x' at $t = 1$, see [Eme89].

In what follows x will be a diffusion on M satisfying by the SDE

$$dx = A_0(x)dt + A_m(x) \circ dw^m, \tag{3.12}$$

where $A_0, A_1, \ldots, A_d$ are vector fields on M and w is a standard Brownian motion in $\mathbf{R}^d$. Let

$$\pi : 0 = t_0 < t_1 < t_2 < \cdots$$

be a division of the time interval $[0, \infty]$ with diameter

$$|\pi| = \sup_{i=1,2,\ldots} |t_i - t_{i-1}| .$$

We define a *piecewise geodesic approximation* of x by

$$x_\pi(t) = I\left(x(t_{i-1}), x(t_i), \frac{t - t_{i-1}}{t_i - t_{i-1}}\right).$$

for any $t \in [t_{i-1}, t_i]$ and $i = 1, 2, \ldots$. The sample paths of x_π are continuous piecewise geodesic curves on M. Each curve of this kind can be lifted to TM by taking the horizontal lift of each smooth segment and pasting the resulting pieces together into a continuous piecewise smooth curve y_π on TM. Each smooth segment of y_π satisfies the differential equation

$$\frac{dy_\pi^i}{dt} = -\gamma_{jk}^i(x_\pi) y_\pi^j \frac{dx_\pi^k}{dt}.$$

The following result is well known, see [Eme89].

Proposition 3.2. *As $|\pi| \to 0$ the approximation (x_π, y_π) tends in probability uniformly with respect to t in any compact set to a diffusion (x, y) on TM satisfying the SDE*

$$dy^i = -\gamma_{jk}^i(x) y^j \circ dx^k. \tag{3.13}$$

This SDE can be expressed in coordinate-free form as

$$dy = h(A_0)(x, y)dt + h(A_m)(x, y) \circ dw^m. \tag{3.14}$$

If (x, y) is a diffusion on TM satisfying (3.13), then y is called a *horizontal lift* of x from M to TM.

Proposition 3.1 can be extended to diffusions as follows.

Proposition 3.3. *Let x be a diffusion on M defined by (3.12). If y and $\tilde{y}$ are horizontal lifts of x to TM, then*

$$g_{ij}(x) y^i \tilde{y}^j = \text{const} \quad P\text{-a.s.}$$

Proof. Compute the stochastic differential $d\left(g_{ij}(x)y^i\tilde{y}^j\right)$ with the aid of the Itô formula, using (3.13) to express dy^i and $d\tilde{y}^j$ and applying the metricity condition (3.6) of the Levi-Civita connection to get

$$\begin{aligned} d\left(g_{ij}(x)y^i\tilde{y}^j\right) &= (\partial_k g_{ij})y^i\tilde{y}^j \circ dx^k + g_{ij}\tilde{y}^j \circ dy^i + g_{ij}y^i \circ d\tilde{y}^j \\ &= \left(\partial_k g_{ij} - \gamma_{kj}^n g_{in} - \gamma_{ik}^n g_{nj}\right) y^i\tilde{y}^j \circ dx^k \\ &= 0. \end{aligned}$$

It follows that $g_{ij}(x)y^i\tilde{y}^j$ is constant P-a.s. ∎

By Theorem 2.6 the horizontal lift y of x to TM exists for any initial condition $y(0) \in TM_{x(0)}$ and it is unique in the sense that if $\tilde{y}$ is also a horizontal lift of x and P-a.s. $y(0) = \tilde{y}(0)$, then P-a.s. $y = \tilde{y}$. Here y and $\tilde{y}$ are understood as maximal solutions of the SDE (3.14). Moreover, Proposition 3.3 implies that the length of y is constant P-a.s., so by Theorem 2.7 the explosion time of y must be the same as that of x.

The mapping

$$\Pi_{s,t} : TM_{x(s)} \ni v \mapsto y(s,v;t) \in TM_{x(t)},$$

where $y(s,v;\cdot)$ is the horizontal lift of x to TM such that $y(s,v;s) = v$, is called the *stochastic parallel transport* along x. Proposition 3.3 implies that $\Pi_{s,t}$ is P-a.s. a unitary mapping from $TM_{x(s)}$ to $TM_{x(t)}$.

A diffusion x on M can also be lifted to the orthonormal frame bundle OM in much the same way as a smooth function can. For a diffusion (x,z) on OM we say that $z = (z_1, \ldots, z_d)$ is a *horizontal lift* of x if to OM if each TM-valued diffusion z_m is a horizontal lift of x to TM, that is,

$$dz_n = h(A_0)(x,z_n)dt + h(A_m)(x,z_n) \circ dw^m$$

for each $n = 1, \ldots, d$. An equivalent way of writing this condition is

$$dz = l(A_0)(x,z)dt + l(A_m)(x,z) \circ dw^m$$

or, in local coordinates,

$$dz_n^i = -\gamma_{jk}^i(x) z_n^j \circ dx^k. \tag{3.15}$$

These equations should be compared with (3.7), (3.8) and (3.9).

Theorems 2.6, 2.7 and Proposition 3.3 ensure that z exists and is unique to within P-a.s. equality for any initial condition $z(0) \in OM_{x(0)}$ and that the explosion time of z is the same as that of x.

We can also define stochastic versions of the horizontal lift of a diffusion to any tensor bundle $T^n_m M$ and the parallel transport of tensors along a diffusion. Namely, for a diffusion (x, u) on $T^n_m M$ we say that u is a *horizontal lift* of a diffusion x from M to $T^n_m M$ if for any horizontal lift z of x to OM the components of $u(t)$ in the frame $z(t)$ are P-a.s. constant. The *stochastic parallel transport of tensors* from $T^n_m M$ along x is defined by

$$\Pi_{s,t} : T^n_m M_{x(s)} \ni v \mapsto u(s, v; t) \in T^n_m M_{x(t)},$$

where $u(s, v; \cdot)$ is the horizontal lift of x to $T^n_m M$ with initial condition $u(s, v; s) = v$.

Let M and $\tilde{M}$ be d-dimensional Riemannian manifolds. The notion of Riemannian rolling can be extended to the case of 'rolling' the manifolds along diffusions x and $\tilde{x}$ on M and $\tilde{M}$, respectively. Here we assume that x is given by the SDE (3.12) with vector fields $A_0, A_1, \ldots, A_d$ and $\tilde{x}$ by a similar SDE with vector fields $\tilde{A}_0, \tilde{A}_1, \ldots, \tilde{A}_d$. We take z and $\tilde{z}$ to be horizontal lifts of x and $\tilde{x}$ to OM and $O\tilde{M}$, respectively. Then the no-slipping condition (3.10) can be generalized as follows:

$$\zeta^i_j \circ dx^j = \tilde{\zeta}^i_j \circ d\tilde{x}^j, \tag{3.16}$$

where ζ and $\tilde{\zeta}$ are the dual frames to z and $\tilde{z}$. If the initial conditions $x(0), z(0)$ and $\tilde{x}(0), \tilde{z}(0)$ are fixed, then the system of SDEs (3.16) defines a one-to-one mapping between diffusions x and $\tilde{x}$ on the two manifolds. This mapping is called the (*Riemannian*) *stochastic rolling.* It is independent of the choice of z and $\tilde{z}$ as long as (3.16) is satisfied.

In particular, in the most typical situation when $\tilde{M} = \mathbb{R}^d$ and $\tilde{x} = w$ is a Brownian motion in $\mathbb{R}^d$ equations (3.16) and (3.15) reduce to

$$\begin{aligned} dx^i &= z^i_n \circ dw^n, && \text{(3.17a)} \\ dz^i_n &= -\gamma^i_{jk}(x) z^j_n z^k_m \circ dw^m, && \text{(3.17b)} \end{aligned}$$

see (3.11). For any initial conditions $x(0), z(0)$ the system of SDEs (3.17) defines a unique (to within equality P-a.s.) diffusion (x, z) on OM, which is called the (*Riemannian*) *stochastic development.*

It turns out that x has the Markov property (2.1) and the probability law of x depends only on $x(0)$, but not on $z(0)$, see [IW89], [Elw82] or [Eme89].

From (3.17) we find that

$$\begin{aligned} dx^i &= z^i_n dw^n + \frac{1}{2} dz^i_n dw^n \\ &= z^i_n dw^n - \frac{1}{2}\gamma^i_{jk} z^j_n z^k_m dw^m dw^n \\ &= z^i_n dw^n - \frac{1}{2}\gamma^i_{jk} g^{jk} dt, \\ dx^i dx^j &= z^i_n z^j_m dw^n dw^m = g^{ij} dt, \end{aligned}$$

since $dw^m dw^n = \delta^{mn} dt$ and $z^i_n z^j_m \delta^{mn} = g^{ij}$. Taking any smooth function f from M to $\mathbb{R}$ and applying the Itô formula, we compute

$$\begin{aligned} df(x) &= (\partial_i f) dx^i + \frac{1}{2}(\partial_i \partial_j f) dx^i dx^j \\ &= (\partial_i f) z^i_n dw^n + \frac{1}{2} g^{ij} \left(\partial_i \partial_j f - \gamma^k_{ij}(x) \partial_k f\right) dt. \end{aligned}$$

It follows that the differential generator of x is

$$D = \frac{1}{2}\Delta = \frac{1}{2} g^{ij} \left(\partial_i \partial_j f - \gamma^k_{ij}(x) \partial_k f\right),$$

where Δ the Laplace-Beltrami operator. A diffusion x on M with differential generator $\frac{1}{2}\Delta$ is called a *Brownian motion* on M.

3.2 Rolling Finsler Manifolds Along Smooth Curves and Diffusions

From now on M will be a d-dimensional Finsler manifold with metric function $L(x, y)$, metric tensor $g_{ij}(x, y)$ and the Cartan connection with coefficients $\left(\mathbf{N}^i_j, \mathbf{F}^i_{jk}, \mathbf{C}^i_{jk}\right)$.

3.2.1 Deterministic Case

On a Finsler manifold the notion of parallelism along a curve splits into two alternative concepts, namely, h-parallelism and hv-parallelism together with the corresponding horizontal lifts, rolling and development. To begin with, we shall introduce these concepts in the deterministic setting. This will then be used to set up an approximation procedure, which gives rise to stochastic h- and hv-horizontal lifts and rolling along a diffusion.

h-Rolling

Let X be a Finsler vector field on M. (In particular, X may be an ordinary vector field on M.) We shall define the *n-horizontal lift* $h(X)$ and the corresponding n-horizontal lift of a smooth curve from M to $\mathring{T}M$. These will be determined by the nonlinear connection $\mathbf{N}_i^j$. Namely, $h(X)$ is a vector field on $\mathring{T}M$ defined by

$$h(X)(x,y) = X^i(x,y)\left(\partial_i - \mathbf{N}_i^j(x,y)\dot{\partial}_j\right). \tag{3.18}$$

Let x be an integral curve of a vector field X on M, that is $\frac{dx}{dt} = X$. For a smooth curve (x,y) on TM we say that y is an *n-horizontal lift* of x to $\mathring{T}M$ if

$$\frac{dy}{dt} = h(X)(x,y),$$

which in local coordinates takes the form

$$\frac{dy^i}{dt} = -\mathbf{N}_j^i(x,y)\frac{dx^j}{dt} \tag{3.19}$$

or, equivalently,

$$\frac{\delta y^i}{dt} = 0. \tag{3.20}$$

This defines y uniquely for any initial condition $y(0) \in \mathring{T}M_{x(0)}$.

Proposition 3.4. *If y is an n-horizontal lift of a smooth curve x from M to $\mathring{T}M$, then*

$$L(x,y) = \text{const}.$$

Proof. First we compute

$$\begin{aligned}
\delta_i L^2 &= \delta_i\left(g_{jk}y^j y^k\right) \\
&= \left(\delta_i g_{jk}\right) y^j y^k + g_{jk}\left(\delta_i y^j\right) y^k + g_{jk} y^j \left(\delta_i y^k\right) \\
&= \left(\delta_i g_{jk} - g_{jn}\mathbf{F}_{ik}^n - g_{kn}\mathbf{F}_{ij}^n\right) y^j y^k = 0,
\end{aligned}$$

since $\delta_i y^j = -\mathbf{N}_i^j = -\mathbf{F}_{ik}^j y^k$ (the Cartan connection is deflection-free) and $\delta_i g_{jk} = g_{jn}\mathbf{F}_{ik}^n + g_{kn}\mathbf{F}_{ij}^n$ (the Cartan connection is h-metrical). But $\delta_i L^2 = 2L\delta_i L$, so

$$\delta_i L = 0.$$

Thus, by (3.20) we have

$$\frac{d}{dt}L(x,y) = \delta_i L(x,y)\frac{dx^i}{dt} + \dot{\partial}_i L(x,y)\frac{\delta y^i}{dt} = 0,$$

which proves the proposition. ∎

The *n-parallel transport* along x is defined by

$$\Pi^h_{s,t} : \mathring{T}M_{x(s)} \ni v \mapsto y(s,v;t) \in \mathring{T}M_{x(t)},$$

where $y(s,v;\cdot)$ is the n-horizontal lift of x to $\mathring{T}M$ with initial condition $y(s,v;s) = v$.

The *h-horizontal lift* $l(X)$ of a Finsler vector field X to OM is a vector field on the Finsler orthonormal frame bundle OM defined by

$$l(X)(x,y,z) = X^i(x,y)\left(\delta_i - \mathbf{F}^k_{ij}(x,y)z^j_n\partial^n_k\right).$$

We say that a smooth curve z on OM is an *h-horizontal lift* of a smooth curve x on M if

$$\frac{dz}{dt} = l(X)(x,y,z),$$

where X is a vector field on M such that x is an integral curve of X and y is an n-horizontal lift of x to $\mathring{T}M$. In local coordinates this differential equation can be written as

$$\frac{dz^i_n}{dt} = -\mathbf{F}^i_{jk}(x,y)z^j_n\frac{dx^k}{dt}, \tag{3.21}$$

where y satisfies (3.20). This defines z uniquely for any initial conditions $z(0) \in OM_{x(0)}$ and $y(0) \in \mathring{T}M_{x(0)}$.

The *h-horizontal lift* of any smooth curve x from M to T^n_mM can be defined as follows. Let (x,u) be a smooth curve on T^n_mM such that for any h-horizontal lift z of x to OM the components of u are constant in the frame z. Such a u is defined uniquely for any initial conditions $u(0) \in T^n_mM_{x(0)}$ and $y(0) \in \mathring{T}M_{x(0)}$. In particular, in $T^1_0M = TM$ the h-horizontal lift u satisfies the differential equation

$$\frac{du^i}{dt} = -\mathbf{F}^i_{jk}(x,y)u^j\frac{dx^k}{dt},$$

where y satisfies (3.20).

Because the Cartan connection is deflection-free, if y is an n-horizontal lift of x, then it is also an h-horizontal lift of x, but not the other way round. On a Riemannian manifold both the n-horizontal lift and the h-horizontal lift of x reduce to the differential equation (3.5), so they are the same.

Let M and $\tilde{M}$ be two d-dimensional Finsler manifolds. We take smooth curves x and $\tilde{x}$ on M and $\tilde{M}$ and their h-horizontal lifts z and $\tilde{z}$ to OM and $O\tilde{M}$, respectively. Thus, z satisfies (3.21), where y is a solution of (3.19). Similarly, $\tilde{z}$ and $\tilde{y}$ satisfy (3.21) and (3.19) with $x, y, \mathbf{F}^i_{jk}, \mathbf{N}^i_j$ replaced by $\tilde{x}, \tilde{y}, \tilde{\mathbf{F}}^i_{jk}, \tilde{\mathbf{N}}^i_j$. We shall say that the manifolds *h-roll* against one another without slipping if

$$\zeta^i_j \frac{dx^j}{dt} = \tilde{\zeta}^i_j \frac{d\tilde{x}^j}{dt},$$

where ζ and $\tilde{\zeta}$ are the dual frames to z and $\tilde{z}$, just like in the Riemannian case, see (3.10). For initial conditions $x(0), y(0), z(0)$ and $\tilde{x}(0), \tilde{y}(0), \tilde{z}(0)$ this defines a one-to-one mapping between smooth curves x and $\tilde{x}$ on M and $\tilde{M}$, which will be called the *Finslerian h-rolling*.

hv-Rolling

The *hv-horizontal lift* $l(Y)$ of a vector field $Y = X^i\delta_i + Y^i\dot{\partial}_i$ on $\mathring{T}M$ is a vector field on the Finsler orthonormal frame bundle OM defined by

$$\begin{aligned} l(Y)(x, y, z) &= X^i(x, y)\left(\delta_i - \mathbf{F}^k_{ij}(x, y) z^j_n \partial^n_k\right) \\ &\quad + Y^i(x, y)\left(\dot{\partial}_i - \mathbf{C}^k_{ij}(x, y) z^j_n \partial^n_k\right), \end{aligned} \tag{3.22}$$

see Sections 1.13 and 1.14. Let (x, y) be a smooth curve on $\mathring{T}M$ and let (x, y, z) be a smooth curve on OM. We say that z is an *hv-horizontal lift* of (x, y) if

$$\frac{dz}{dt} = l(Y)(x, y, z),$$

where Y is a vector field on $\mathring{T}M$ such that (x, y) is an integral curve of Y. In local coordinates this differential equation can be written as follows:

$$\frac{dz^i_n}{dt} = -\mathbf{F}^i_{jk}(x, y) z^j_n \frac{dx^k}{dt} - \mathbf{C}^i_{jk}(x, y) z^j_n \frac{\delta y^k}{dt}. \tag{3.23}$$

This defines z uniquely for any initial condition $z(0) \in OM_{(x(0), y(0))}$.

Now the *hv*-horizontal lift of any smooth curve (x, y) from $\mathring{T}M$ to $T^n_m M$ can be defined to be a smooth curve (x, u) on $T^n_m M$ such that for any *hv*-horizontal lift z of y to OM the components of u are constant in the frame z. This definition does not depend on the choice of z.

In particular, if u is a *hv*-horizontal lift of (x, y) to $T^1_0 M = TM$, than it satisfies the differential equation

$$\frac{du^i}{dt} = -\mathbf{F}^i_{jk}(x, y) u^j \frac{dx^k}{dt} - \mathbf{C}^i_{jk}(x, y) u^j \frac{\delta y^k}{dt}. \tag{3.24}$$

When M is a Riemannian manifold, then $\mathbf{C}^i_{jk}(x, y) = 0$ and $\mathbf{F}^i_{jk}(x, y) = \gamma^i_{jk}(x)$, so the *hv*-horizontal lift of (x, y) reduces to the horizontal lift of x as defined in Section 3.1.1.

Proposition 3.5. *Let (x, y) be a smooth curve on $\mathring{T}M$. Suppose that u and v are hv-horizontal lifts of y to $T^1_0 M = TM$. Then*

$$g_{ij}(x, y) u^i v^j = \text{const}\,.$$

Proof. Differentiate $g_{ij}(x, y) u^i v^j$ with respect to t, use (3.24) to express $\frac{du^i}{dt}$ and $\frac{dv^i}{dt}$, and apply the h- and v-metricity conditions (1.88) of the Cartan connection to show that

$$\frac{d}{dt} g_{ij}(x, y) u^i v^j = 0,$$

which proves the proposition. ∎

Let M and $\tilde{M}$ be two d-dimensional Finsler manifolds. We take smooth curves (x, y) and $(\tilde{x}, \tilde{y})$ on $\mathring{T}M$ and $\mathring{T}\tilde{M}$, and let z and $\tilde{z}$ be their *hv*-horizontal lifts to OM and $O\tilde{M}$, respectively. This means that z satisfies equation (3.23) and $\tilde{z}$ satisfies the same equation with $x, y, \mathbf{F}^i_{jk}, \mathbf{C}^i_{jk}$ and δ replaced by $\tilde{x}, \tilde{y}, \tilde{\mathbf{F}}^i_{jk}, \tilde{\mathbf{C}}^i_{jk}$ and $\tilde{\delta}$. We shall say that the manifolds *hv-roll* against one another without slipping if

$$\zeta^i_j \frac{dx^j}{dt} = \tilde{\zeta}^i_j \frac{d\tilde{x}^j}{dt},$$

where ζ and $\tilde{\zeta}$ are the dual frames to z and $\tilde{z}$. Given the initial conditions $x(0), y(0), z(0)$ and $\tilde{x}(0), \tilde{y}(0), \tilde{z}(0)$ this defines a one-to-one mapping between smooth curves (x, y) and $(\tilde{x}, \tilde{y})$ on $\mathring{T}M$ and $\mathring{T}\tilde{M}$, which will be called the *Finslerian hv-rolling*.

3.2.2 Stochastic h-Rolling of Finsler Spaces

In Section 3.2.1 we discussed the n-horizontal and h-horizontal lifts in the deterministic setting. One of the goals of the present section is to extend these notions to the case of diffusions on a Finsler manifold. We shall proceed in a similar manner as in Section 3.1.2, that is, we shall approximate diffusions by processes with piecewise smooth sample paths and investigate the limit of the corresponding lifts.

An interpolation rule on a Finsler manifold is defined by the same conditions (I1)–(I3) as on a Riemannian manifold in Section 3.1.2. We shall use the *geodesic interpolation rule* I *on* M such that $t \mapsto I(x, x', t)$ is the shortest geodesic with $I(x, x', 0) = x$ and $I(x, x', 1) = x'$. Proposition (7.13) in [Eme89] shows in detail how to verify that conditions (I1)–(I3) are indeed satisfied in this case.

We take x to be a diffusion on a Finsler manifold M such that

$$dx = A_0(x)dt + A_m(x) \circ dw^m, \tag{3.25}$$

where $A_0, A_1, \dots, A_d$ are vector fields on M, and w is a standard Brownian motion in $\mathbb{R}^d$. We define a *piecewise geodesic approximation* of x by

$$x_\pi(t) = I\left(x(t_{i-1}), x(t_i), \frac{t - t_{i-1}}{t_i - t_{i-1}}\right)$$

for any $t \in [t_{i-1}, t_i]$ and $i = 1, \dots, n$, where

$$\pi : 0 = t_0 < t_1 < t_2 < \cdots$$

is a division of $[0, \infty]$. The sample paths of x_π are continuous piecewise geodesic curves on M. Each curve of this kind can be lifted to TM by taking the n-horizontal lift of each smooth segment and pasting the resulting pieces together into a continuous piecewise smooth curve y_π on TM. Each smooth piece of y_π satisfies the differential equation

$$\frac{dy_\pi^i}{dt} = -\mathbf{N}_j^i(x_\pi, y_\pi)\frac{dx_\pi^j}{dt}.$$

The standard theorem on approximating solutions of Stratonovich SDE's, see e.g. Theorem (7.24) in [Eme89], implies the following result.

Theorem 3.6. *As $|\pi| \to 0$ the approximation (x_π, y_π) tends in probability uniformly with respect to t in any compact set to a diffusion (x, y) on $\mathring{T}M$ projecting onto x and satisfying the SDE*

$$dy^i = -\mathbf{N}_j^i(x, y) \circ dx^k. \tag{3.26}$$

This SDE can be expressed in coordinate-free form as

$$dy = h(A_0)(x,y)dt + h(A_m)(x,y) \circ dw^m, \tag{3.27}$$

where h is defined by (3.18).

A diffusion (x,y) on $\mathring{T}M$ such that y satisfies this SDE is called an *n-horizontal lift* of x from M to $\mathring{T}M$.

For a diffusion (x,y) on $\mathring{T}M$ we introduce the notation

$$\delta y^j = dy^i + \mathbf{N}^i_j(x,y) \circ dx^k. \tag{3.28}$$

Thus, for example, equation (3.26) can be written as (see (3.20))

$$\delta y^i = 0. \tag{3.29}$$

Proposition 3.7. *If x is a diffusion on M and y is an n-horizontal lift of x to $\mathring{T}M$, then*

$$L(x,y) = \text{const} \qquad P\text{-a.s.}$$

Proof. It was shown in Proposition 3.4 that $\delta_i L(x,y) = 0$. Thus, computing the stochastic differential of $L(x,y)$, we obtain

$$\begin{aligned} dL(x,y) &= \partial_i L(x,y) \circ dx^i + \dot{\partial}_i L(x,y) \circ dy^i \\ &= \delta_i L(x,y) \circ dx^i + \dot{\partial}_i L(x,y) \circ \delta y^i = 0. \end{aligned}$$

by the Itô formula and (3.29). This proves that $L(x,y) = \text{const}$ P-a.s. ∎

By Theorem 2.6 the n-horizontal lift y of x to $\mathring{T}M$ exists for any initial condition $y(0) \in TM_{x(0)}$ and it is unique in the sense that if $\tilde{y}$ is also a horizontal lift of x and P-a.s. $y(0) = \tilde{y}(0)$, then P-a.s. $y = \tilde{y}$. Here y and $\tilde{y}$ are understood as maximal solutions of the SDE (3.27). Moreover, Proposition 3.7 implies that the length $L(x,y)$ of y is constant P-a.s., so by Theorem 2.7 the explosion time of y must be the same as that of x.

Let (x,y,z) be a diffusion on the Finsler orthonormal frame bundle OM. We say that z is an *h-horizontal lift* of x to OM if

$$dz^i_n = -\mathbf{F}^i_{jk}(x,y) z^j_n \circ dx^k, \tag{3.30}$$

where y is an n-horizontal lift of x to $\mathring{T}M$. This SDE can be written in the coordinate-free form

$$dz = l(A_0)(x, y, z)dt + l(A_m)(x, y, z) \circ dw^m,$$

where $A_0, A_1, \ldots, A_n$ are the vector fields defining x in (3.25). In this way z is defined uniquely for any initial conditions $y(0) \in \mathring{T}M_{x(0)}$ and $z(0) \in OM_{(x(0),y(0))}$.

The *h-horizontal lift* of a diffusion x from M to $T^n_m M$ can be defined as a diffusion (x, u) on $T^n_m M$ such that for any h-horizontal lift z of x to OM the components of u are constant in the frame z. Such a u is defined uniquely for any initial conditions $u(0) \in T^n_m M_{x(0)}$ and $y(0) \in \mathring{T}M_{x(0)}$. In particular, in $T^1_0 M = TM$ the h-horizontal lift u satisfies the SDE

$$du^i = -\mathbf{F}^i_{jk}(x, y)u^j \circ dx^k,$$

where y satisfies (3.26).

As in the deterministic case, if y is an n-horizontal lift of x, then it is also an h-horizontal lift of x, but not the other way round. This is because the Cartan connection is deflection-free. On a Riemannian manifold the n-horizontal lift and the h-horizontal lift of x both reduce to the SDE (3.13), so they are the same.

Let M and $\tilde{M}$ be two d-dimensional Finsler manifolds. We take diffusions x and $\tilde{x}$ on M and $\tilde{M}$ and their h-horizontal lifts z and $\tilde{z}$ to OM and $O\tilde{M}$, respectively. Thus, z satisfies (3.30), where y is a solution of (3.26). Similarly, $\tilde{z}$ and $\tilde{y}$ satisfy (3.30) and (3.26) with $x, y, \mathbf{F}^i_{jk}, \mathbf{N}^i_j$ replaced by $\tilde{x}, \tilde{y}, \tilde{\mathbf{F}}^i_{jk}, \tilde{\mathbf{N}}^i_j$. We shall say that the manifolds *h-roll* against one another without slipping if

$$\zeta^i_j \circ dx^j = \tilde{\zeta}^i_j \circ d\tilde{x}^j, \tag{3.31}$$

where ζ and $\tilde{\zeta}$ are the dual frames to z and $\tilde{z}$. Given the initial conditions $x(0), y(0), z(0)$ and $\tilde{x}(0), \tilde{y}(0), \tilde{z}(0)$ this defines a one-to-one mapping between diffusions x and $\tilde{x}$ on M and $\tilde{M}$, which will be called the *Finslerian stochastic h-rolling.*

3.2.3 Stochastic *hv*-Rolling of Finsler Spaces

To generalize the hv-horizontal lift we shall approximate a diffusion (x, y) on $\mathring{T}M$ (rather than on M, as in the previous section) by a process with continuous piecewise smooth sample paths. To this end we need an interpolation rule on $\mathring{T}M$ rather than on M. Such an interpolation rule can be obtained by introducing the so-called *diagonal lift metric*

$$G(x,y) = g_{ij}(x,y)dx^i \otimes dx^j + g_{ij}(x,y)\delta y^i \otimes \delta y^j.$$

Since the Finsler metric tensor g is assumed to be non-degenerate and positive definite, this defines a Riemannian metric on $\mathring{T}M$ with metric tensor G. Now we are in a position to consider the *geodesic interpolation rule* J *on* $\mathring{T}M$ such that $t \mapsto J((x,y),(x',y'),t)$ is a geodesic on the Riemannian manifold $\mathring{T}M$ with the diagonal lift metric G for any $x, x' \in M$ and $y \in \mathring{T}M_x$, $y' \in \mathring{T}M_{x'}$. That this is indeed an interpolation rule follows directly from Proposition (7.13) in [Eme89].

Let (x,y), where $y \in \mathring{T}M_x$, be a diffusion on $\mathring{T}M$ satisfying the SDE

$$d(x,y) = Y_0(x,y)dt + Y_m(x,y) \circ dw^m, \tag{3.32}$$

where $Y_0, Y_1, \ldots, Y_d$ are vector fields on $\mathring{T}M$. Writing $Y_m = A_m\delta_i + B_m\dot{\partial}_i$ for $m = 0, 1, \ldots, d$, where the A_m and B_m are Finsler vector fields on M, we can express (3.32) in local coordinates as

$$\begin{aligned} dx^i &= A_0^i(x,y)dt + A_m^i(x,y) \circ dw^m, \\ \delta y^i &= B_0^i(x,y)dt + B_m^i(x,y) \circ dw^m. \end{aligned}$$

We define a *piecewise geodesic approximation* of (x,y) by

$$(x_\pi(t), y_\pi(t)) = J\left((x(t_{i-1}), y(t_{i-1})), (x(t_i), y(t_i)), \frac{t - t_{i-1}}{t_i - t_{i-1}} \right)$$

for any $t \in [t_{i-1}, t_i]$ and $i = 1, \ldots, n$, and any division

$$\pi : 0 = t_0 < t_1 < t_2 < \cdots .$$

The sample paths of (x_π, y_π) are continuous piecewise geodesic curves on $\mathring{T}M$. Now we can take the hv-horizontal lift of each sample path of (x_π, y_π), obtaining a piecewise smooth process (x_π, y_π, z_π) on the Finsler orthonormal frame bundle OM. (To be precise, we lift each smooth segment of each piecewise smooth path and then paste the lifted segments together to obtain continuous piecewise smooth paths on OM.) Each smooth piece of the lifted process satisfies the differential equation

$$\frac{d(z_\pi)_n^i}{dt} = -\mathbf{F}^i_{jk}(x_\pi, y_\pi)(z_\pi)_n^j \frac{dx_\pi^k}{dt} - \mathbf{C}^i_{jk}(x_\pi, y_\pi)(z_\pi)_n^j \frac{\delta y_\pi^k}{dt}.$$

From Theorem (7.24) in [Eme89] we immediately obtain the following result.

Theorem 3.8. *As $|\pi| \to 0$ the approximation (x_π, y_π, z_π) tends in probability uniformly with respect to t in any compact set to a diffusion (x, y, z) on OM satisfying the SDE*

$$dz_n^i = -\mathbf{F}^i_{jk}(x,y) z_n^j \circ dx^k - \mathbf{C}^i_{jk}(x,y) z_n^j \circ \delta y^k. \tag{3.33}$$

This SDE can be written in coordinate-free form as

$$dz = l(Y_0)(x,y,z)dt + l(Y_m)(x,y,z) \circ dw^m, \tag{3.34}$$

where l is the hv-horizontal lift (3.22).

We say that a diffusion z on OM satisfying (3.34) is an *hv-horizontal lift* of (x, y). Theorem 2.6 implies that z exists and is unique for any initial condition $z(0) \in OM_{(x(0),y(0))}$. Because the fibres of OM are compact, the explosion time of z cannot be less than that of (x, y).

In a similar way as in the previous sections we can define an hv-horizontal lift of a diffusion (x, y) from $\mathring{T}M$ to any tensor bundle $T^n_m M$. For a diffusion (x, u) on $T^n_m M$ we say that u is an *hv-horizontal lift* of (x, y) from M to $T^n_m M$ if for any hv-horizontal lift z of (x, y) to OM the components of u in the frame z are P-a.s. constant. Such a diffusion u is defined uniquely (to within equality P-a.s.) for any initial condition $u(0) \in T^n_m M$ and is independent of the choice of z. I particular, if u is an hv-horizontal lift of (x, y) to $T^1_0 M = TM$, then it satisfies the SDE

$$du^i = -\mathbf{F}^i_{jk}(x,y) u^j \circ dx^k - \mathbf{C}^i_{jk}(x,y) u^j \circ \delta y^k. \tag{3.35}$$

The *stochastic hv-parallel transport of tensors* from $T^n_m M$ along (x, y) is defined by

$$\Pi_{s,t} : T^n_m M_{x(s)} \ni v \mapsto u(s, v; t) \in T^n_m M_{x(t)},$$

where $u(s, v; \cdot)$ is the hv-horizontal lift of (x, y) to $T^n_m M$ with initial condition $u(s, v; s) = v$.

Proposition 3.9. *Let (x, y) be a diffusion on $\mathring{T}M$. If u and v are hv-horizontal lifts of (x, y) to TM, then*

$$g_{ij}(x,y) u^i v^j = \text{const} \qquad P\text{-a.s.}$$

Proof. Applying the Itô formula and using (3.35) to express du^i and dv^j, we compute the stochastic differential

$$\begin{aligned} d\left(g_{ij}(x,y)u^i v^j\right) &= (\delta_k g_{ij}) u^i v^j \circ dx^k + (\dot{\partial}_k g_{ij}) u^i v^j \circ \delta y^k \\ &\quad + g_{ij} v^j \circ du^i + g_{ij} u^i \circ dv^j \\ &= (\delta_k g_{ij} - g_{in}\mathbf{F}^n_{kj} - g_{nj}\mathbf{F}^n_{ki}) u^i v^j \circ dx^k \\ &\quad + (\dot{\partial}_k g_{ij} - g_{in}\mathbf{C}^n_{kj} - g_{nj}\mathbf{C}^n_{ki}) u^i v^j \circ \delta y^k = 0, \end{aligned}$$

which is equal to zero because of the h- and v-metricity conditions (1.88) of the Cartan connection. It follows that P-a.s. $g_{ij}(x,y)u^i v^j = \text{const}$, as required. ∎

Let M and $\tilde{M}$ be two d-dimensional Finsler manifolds. We take diffusions (x,y) and $(\tilde{x},\tilde{y})$ on $\mathring{T}M$ and $\mathring{T}\tilde{M}$. Let z and $\tilde{z}$ be their hv-horizontal lifts to OM and $O\tilde{M}$, respectively. This means that z satisfies the SDE (3.33) and $\tilde{z}$ satisfies the same equation with $x, y, \mathbf{F}^i_{jk}, \mathbf{C}^i_{jk}$ and δ replaced by $\tilde{x}, \tilde{y}, \tilde{\mathbf{F}}^i_{jk}, \tilde{\mathbf{C}}^i_{jk}$ and $\tilde{\delta}$.

We shall say that the manifolds *hv-roll* against one another without slipping if

$$\zeta^i_j \circ dx^j = \tilde{\zeta}^i_j \circ d\tilde{x}^j, \tag{3.36}$$

$$\zeta^i_j \circ \delta y^j = \tilde{\zeta}^i_j \circ \tilde{\delta}\tilde{y}^j, \tag{3.37}$$

where ζ and $\tilde{\zeta}$ are the dual frames to z and $\tilde{z}$. Given the initial conditions $x(0), y(0), z(0)$ and $\tilde{x}(0), \tilde{y}(0), \tilde{z}(0)$ this defines a one-to-one mapping between diffusions (x,y) and $(\tilde{x},\tilde{y})$ on $\mathring{T}M$ and $\mathring{T}\tilde{M}$, which will be called the *Finslerian stochastic hv-rolling.*

3.3 Finslerian Stochastic Development

We shall apply the concepts introduced in Section 3.2 to roll a Finsler manifold along a Brownian motion in a Euclidean space. Let M be a d-dimensional Finsler manifold and let $\mathbb{R}^d$ with the ordinary Euclidean metric play the role of another Finsler manifold. Then $\mathbb{R}^d \times \mathbb{R}^d$ can be identified with the tangent space $T\mathbb{R}^d$. We denote by (w,v) a Brownian motion in $\mathbb{R}^d \times \mathbb{R}^d$, that is, w and v will be independent Brownian motions in $\mathbb{R}^d$. According to the discussion in the previous section, M can either be h-rolled along w or hv-rolled along (w,v). This leads to two alternative approaches to Finslerian stochastic development.

In the case in hand equations (3.31), (3.29), (3.30) of Section 3.2.2 reduce to

$$dx^i = z^i_n \circ dw^n, \tag{3.38a}$$

$$\delta y^i = 0, \tag{3.38b}$$

$$dz^i_n = -\mathbf{F}^i_{jk}(x,y) z^j_n z^k_m \circ dw^k. \tag{3.38c}$$

Definition 3.1. A diffusion (x,y,z) on the Finsler orthonormal bundle OM which is a maximal solution of the system of SDEs (3.38) is called a (*Finslerian*) *h-stochastic development.*

On the other hand, equations (3.36), (3.37), (3.33) of Section 3.2.3 reduce to

$$dx^i = z^i_n \circ dw^n, \tag{3.39a}$$
$$\delta y^i = z^i_n \circ dv^n, \tag{3.39b}$$
$$dz^i_n = -\mathbf{F}^i_{jk}(x,y) z^j_n z^k_m \circ dw^m - \mathbf{C}^i_{jk}(x,y) z^j_n z^k_m \circ dv^m. \tag{3.39c}$$

Definition 3.2. A diffusion (x, y, z) on the Finsler orthonormal bundle OM which is a maximal solution of the system of SDEs (3.39) is called a (*Finslerian*) *hv-stochastic development.*

For any initial conditions $(x(0), y(0), z(0)) \in OM$ both systems (3.38) and (3.39) have unique (to within equality P-a.s.) maximal solutions (x, y, z). In general, the solutions are subject to explosions. The results of Section 3.4 will imply that, in fact, only the x part of (x, y, z) can be subject to explosions. In particular, this means that if M is a compact manifold, then (x, y, z) can never develop explosions and is therefore defined for all $t \in [0, \infty)$.

Example 3.1 (Minkowski spaces). A *Minkowski space* M is by definition a finite-dimensional normed vector space whose norm $L(y)$ is smooth for all $y \in \mathring{M} = M \setminus \{0\}$. In this case the slit tangent bundle $\mathring{T}M$ is trivial and can be identified with $M \times \mathring{M}$. A Minkowski space is clearly a Finsler space with metric function

$$M \times \mathring{M} \ni (x, y) \mapsto L(y) \in [0, \infty).$$

It is easily seen that the Cartan connection coefficients $\mathbf{N}^i_j$ and $\mathbf{F}^i_{jk}$ are identically equal to zero, so $\delta_i = \partial_i$, while $\mathbf{C}^i_{jk}(y) = \frac{1}{2} g^{il}(y) \dot{\partial}_l g_{jk}(y)$ and the Finsler metric tensor $g_{ij}(y) = \frac{1}{2} \dot{\partial}_i \dot{\partial}_j L^2(y)$ depend only on $y \in \mathring{T}M \cong \mathring{M}$, but not on $x \in M$. Here we assume in addition that $g_{ij}(y)$ is non-degenerate and positive definite.

In this case the system (3.38) of SDEs for the h-stochastic development takes the form

$$dx^i = z^i_n \circ dw^n,$$
$$dy^i = 0,$$
$$dz^i_n = 0,$$

so $y = \text{const}$ and $z = \text{const}$ P-a.s. and x is just a standard Euclidean Brownian motion on M.

On the other hand, the system (3.39) of SDEs for the hv-stochastic development reduces to

$$\begin{aligned} dx^i &= z^i_n \circ dw^n, \\ dy^i &= z^i_n \circ dv^n, \\ dz^i_n &= -\mathbf{C}^i_{jk}(y) z^j_n z^k_m \circ dv^m. \end{aligned}$$

The last two equations of this system are independent of x and have the same form as the system of SDEs (3.17) for Riemannian stochastic development. Therefore y is a Brownian motion on $\overset{\circ}{M}$ regarded as a Riemannian manifold with metric tensor $g_{ij}(y)$ and Levi-Civita connection $\mathbf{C}^i_{jk}(y)$.

Example 3.2 (Berwald spaces). A *Berwald space* M can be defined as a Finsler space such that $\nabla^h_l \mathbf{C}^i_{jk} = 0$, where $\mathbf{C}^i_{jk}$ are the vertical coefficients of the Cartan connection. As a consequence, there exists a coordinate system in which the horizontal coefficients $\mathbf{F}^i_{jk}$ of the Cartan connection are functions of $x \in M$ only, that is, they are independent of $y \in \overset{\circ}{T}M$, see [Mat86]. In particular, it follows that the nonlinear connection coefficients $\mathbf{N}^i_j$ are in fact linear functions of y, that is, $\mathbf{N}^i_j(x, y) = \mathbf{F}^i_{jk}(x) y^k$ in this coordinate system. Then the SDEs (3.38) for the h-stochastic development take the form

$$\begin{aligned} dx^i &= z^i_n \circ dw^n, \\ \delta y^i &= 0, \\ dz^i_n &= -\mathbf{F}^i_{jk}(x, y) z^j_n z^k_m \circ dw^m. \end{aligned} \tag{3.40}$$

The first and third equations in this system are independent of y. By a theorem of Szabó [Sza81], for any positive definite Berwald space M with compact smooth indicatrix there exists a Riemannian metric $\tilde{g}_{ij}$ on M such that the $\mathbf{F}^i_{jk}$ coincide with the Levi-Civita connection coefficients γ^i_{jk} in this Riemannian metric in the coordinate system in which the $\mathbf{F}^i_{jk}$ depend on $x \in M$ alone. Then the first and third equations in (3.40) are just the SDEs (3.17) for the stochastic development on M regarded as a Riemannian manifold with metric tensor $\tilde{g}_{ij}$, while the second equation in (3.40) means that y is the Riemannian horizontal lift of x. However, the initial conditions for (3.40) corresponding to the Riemannian development and Finslerian h-development are different. Namely, in the Riemannian case the initial fame $z(0)$ must be orthonormal with respect to the Riemannian metric $\tilde{g}_{ij}$, while in the Finslerian case $z(0)$ must be orthonormal with respect to the Finsler metric g_{ij}.

The following results are concerned with the projection of an h- or hv-development (x, y, z) from OM to $\mathring{T}M$.

Theorem 3.10. *If (x, y, z) is a Finslerian h-stochastic development, then the probability law of (x, y) is independent of the initial frame $z(0)$. Moreover, (x, y) satisfies the Markov property and has the differential generator*

$$\begin{aligned} D &= \frac{1}{2}\Delta^h = \frac{1}{2}g^{ij}\nabla_i^h\nabla_j^h \\ &= \frac{1}{2}g^{ij}\left(\delta_i\delta_j - \mathbf{F}_{ij}^k\delta_k\right). \end{aligned}$$

Proof. Let (x, y, z) be a solution to the system of SDEs (3.38) with initial conditions $(x(0), y(0), z(0)) \in OM$. Take any $\tilde{z}(0) \in OM_{x(0),y(0)}$. Then there exists $h \in O(\mathbf{R}^d)$ such that $\tilde{z}_n(0) = h_n^m z_m(0)$. We put $\tilde{z}_n = h_n^m z_m$. It is easily seen that $(x, y, \tilde{z})$ is a solution to (3.38) with w^n replaced by $\tilde{w}^n$ such that $w^m = h_n^m \tilde{w}^n$, and with initial conditions $(x(0), y(0), \tilde{z}(0))$. But $\tilde{w}$ is also a Brownian motion on $\mathbf{R}^d$, so it has the same probability law as w. It follows that the probability law of (x, y) is independent on $z(0)$. Because of this, since (x, y, z) has the Markov property as a solution of the system of SDEs (3.38), it follows (x, y) also has the Markov property.

To compute the generator of (x, y) we express the stochastic differential dx^i in (3.38) and $dx^i dx^j$ as

$$\begin{aligned} dx^i &= z_n^i dw^n + \frac{1}{2}dz_n^i dw^n \\ &= z_n^i dw^n - \frac{1}{2}\mathbf{F}_{jk}^i z_n^j z_m^k dw^m dw^n \\ &= z_n^i dw^n - \frac{1}{2}\mathbf{F}_{jk}^i g^{jk} dt, \\ dx^i dx^j &= z_n^i z_m^j w^n w^m = g^{ij} dt. \end{aligned}$$

Next we evaluate the stochastic differential of any smooth function f on $\mathring{T}M$ with the aid of the Itô formula:

$$\begin{aligned} df(x, y) &= (\delta_i f)dx^i + \frac{1}{2}(\delta_i\delta_j f)dx^i dx^j \\ &= \frac{1}{2}g^{ij}(\delta_i\delta_j f - \mathbf{F}_{ij}^k\delta_k f)dt + (\delta_i f)z_n^i dw^n. \end{aligned}$$

The differential generator $\frac{1}{2}\Delta^h$ is given by the expression multiplying dt above. ∎

Theorem 3.11. *If (x, y, z) is a Finslerian hv-stochastic development, then the probability law of (x, y) is independent of the initial frame $z(0)$. Moreover, (x, y) satisfies the Markov property and has the differential generator*

$$\begin{aligned} D &= \frac{1}{2}\Delta^{hv} = \frac{1}{2}g^{ij}\left(\nabla_i^h\nabla_j^h + \nabla_i^v\nabla_j^v\right) \\ &= \frac{1}{2}g^{ij}\left(\delta_i\delta_j - \mathbf{F}_{ij}^k\delta_k\right) + \frac{1}{2}g^{ij}\left(\dot\partial_i\dot\partial_j - \mathbf{C}_{ij}^k\dot\partial_k\right). \end{aligned}$$

Proof. The proof is similar to that of Theorem 3.10. Let (x, y, z) be a solution to the system of SDEs (3.39) with initial conditions $(x(0), y(0), z(0)) \in OM$. Take any $\tilde z(0) \in OM_{x(0),y(0)}$. Then there exists $h \in O(\mathbf{R}^d)$ such that $\tilde z_n(0) = h_n^m z_m(0)$. We put $\tilde z_n = h_n^m z_m$. It is easily seen that $(x, y, \tilde z)$ is a solution to (3.39) with initial conditions $(x(0), y(0), \tilde z(0))$ and with w^n, v^n replaced by $\tilde w^n, \tilde v^n$, where $w^m = h_n^m \tilde w^n$ and $v^m = h_n^m \tilde v^n$. But $(\tilde w, \tilde v)$ is also a Brownian motion on $\mathbf{R}^d \times \mathbf{R}^d$, so it has the same probability law as (w, v). It follows that the probability law of (x, y) is independent on $z(0)$. Because of this, since (x, y, z) has the Markov property as a solution of the system of SDEs (3.39), it follows (x, y) also has the Markov property.

Next we evaluate the stochastic differentials dx^i and δy^i in (3.39) and their products

$$\begin{aligned} dx^i &= z_n^i dw^n + \frac{1}{2} dz_n^i dw^n \\ &= z_n^i dw^n - \frac{1}{2}\mathbf{F}_{jk}^i z_n^j z_m^k dw^m dw^n - \frac{1}{2}\mathbf{C}_{jk}^i z_n^j z_m^k dv^m dw^n \\ &= z_n^i dw^n - \frac{1}{2}\mathbf{F}_{jk}^i g^{jk} dt \qquad (3.41a) \\ \delta y^i &= z_n^i dv^n + \frac{1}{2} dz_n^i dv^n \\ &= z_n^i dv^n - \frac{1}{2}\mathbf{F}_{jk}^i z_n^j z_m^k dw^m dv^n - \frac{1}{2}\mathbf{C}_{jk}^i z_n^j z_m^k dv^m dv^n \\ &= z_n^i dv^n - \frac{1}{2}\mathbf{C}_{jk}^i g^{jk} dt \qquad (3.41b) \\ dx^i dx^j &= \delta y^i \delta y^j = g^{ij} dt \qquad (3.41c) \\ dx^i \delta y^j &= \delta y^i dx^j = 0. \qquad (3.41d) \end{aligned}$$

We use these expressions to compute the stochastic differential of any smooth function $f : \mathring{T}M \to \mathbf{R}$ with the aid of the Itô formula:

$$\begin{aligned} df(x,y) &= (\delta_i f)dx^i + \frac{1}{2}(\delta_i\delta_j f)dx^i dx^j + \frac{1}{2}(\dot\partial_i\delta_j f)\delta y^i dx^j \\ &\quad +(\dot\partial_i f)\delta y^i + \frac{1}{2}(\delta_i\dot\partial_j f)dx^i\delta y^j + \frac{1}{2}(\dot\partial_i\dot\partial_j f)\delta y^i\delta y^j \\ &= \frac{1}{2}g^{ij}(\delta_i\delta_j f - \mathbf{F}^k_{ij}\delta_k f)dt + (\delta_i f)z^i_n dw^n \\ &\quad +\frac{1}{2}g^{ij}(\dot\partial_i\dot\partial_j f - \mathbf{C}^k_{ij}\dot\partial_k f)dt + (\dot\partial_i f)z^i_n dv^n. \end{aligned}$$

The differential generator $\frac{1}{2}\Delta^{hv}$ is given by the expressions multiplying dt above. ∎

Definition 3.3. A diffusion (x,y) on $\mathring{T}M$ with generator $\frac{1}{2}\Delta^h$ (respectively, $\frac{1}{2}\Delta^{hv}$) is called a Finslerian h-Brownian motion (hv-Brownian motion).

3.4 Radial Behaviour

By the radial behaviour of a diffusion (x,y) on $\mathring{T}M$ we shall understand the behaviour of $L(x,y)$, where L is the Finsler metric function. We shall study the radial behaviour of (x,y) for a stochastic h-development or hv-development (x,y,z). One of the purposes of this study is to ensure that the y part of the development cannot explode. Since the fibres of OM are compact, it is clear that z cannot develop explosions either. Because of this, it is only the x part of Finslerian development that may be subject to explosions.

We observe that an explosion of the y part could *a priori* occur in two ways, namely, y could escape to infinity or to zero in the sense that $L(x,y) \to \infty$ or $L(x,y) \to 0$ as $t \nearrow \tau$, where $\tau < \infty$. If this were the case, it would be impossible to extend the development beyond τ. (In the latter case, when $L(x,y) \to 0$, this is because Finslerian objects, for example, $g_{ij}(x,y)$ are, in general, undefined for $y = 0$.) Therefore we have to ensure that y can neither escape to ∞ nor to 0 in finite time.

Theorem 3.12. a) *If (x,y,z) is a Finslerian h-stochastic development, then*

$$L(x,y) = \text{const} \qquad P\text{-a.s.}$$

b) *If (x,y,z) is a Finslerian hv-stochastic development, then $L(x,y)$ is a* Bessel process *with index* $\dim M$, *that is,*

$$dL^2(x,y) = 2L(x,y)du + (\dim M)dt,$$

where u is a standard Brownian motion in $\mathbb{R}$. *(On the Bessel process see* [IW89], *Example 8.3.)*

Proof. In the proof of Proposition 3.4 it was verified that $\delta_i L^2 = 0$. Moreover,

$$\begin{aligned}\dot\partial_i L^2 &= \dot\partial_i(g_{jk}y^j y^k) = (\dot\partial_i g_{jk})y^j y^k + 2g_{jk}y^j\delta_i^k \qquad (3.42)\\ &= 2\mathbf{C}_{ijk}y^i y^j + 2y_i = 2y_i\end{aligned}$$

by (1.92). Thus, by the Itô formula

$$\begin{aligned}dL^2(x,y) &= \delta_i L^2(x,y)\circ dx^i + \dot\partial_i L^2(x,y)\circ\delta y^i\\ &= \dot\partial_i L^2(x,y)\circ\delta y^i\\ &= \dot\partial_i L^2(x,y)\delta y^i + \frac{1}{2}\delta_i\dot\partial_j L^2(x,y)dx^i\delta y^j + \frac{1}{2}\dot\partial_i\dot\partial_j L^2(x,y)\delta y^i\delta y^j\\ &= 2y_i\delta y^i + g_{ij}(x,y)\delta y^i\delta y^j. \qquad (3.43)\end{aligned}$$

Here we use the fact that $dx^i\delta y^j = 0$ for both the h- and hv-development.

a) Since $\delta y^i = 0$ by (3.38b), it follows from (3.43) that $dL^2(x,y) = 0$. This means that $L^2(x,y) = \text{const}$ P-a.s., which proves assertion a) of the theorem.

b) Because $\delta y^i = z^i_n dv^n - \frac{1}{2}\mathbf{C}^i_{jk}g^{jk}dt$ and $\delta y^i\delta y^j = g^{ij}dt$ by (3.41b) and (3.41c), it follows from (3.43) that

$$\begin{aligned}dL^2(x,y) &= 2y_i\delta y^i + g_{ij}(x,y)\delta y^i\delta y^j\\ &= 2y_i z^i_n dv^n + g_{ij}(x,y)g^{ij}(x,y)dt\\ &= 2L(x,y)\dot\partial_i L(x,y)z^i_n dv^n + (\dim M)\,dt,\end{aligned}$$

since $y_i\mathbf{C}^i_{jk} = 0$ and $2y_i = 2L\dot\partial_i L$. Putting

$$N_n = \dot\partial_i L(x,y)z^i_n$$

and defining u to be an $\mathbb{R}$-valued such that

$$du = N_n dv^n,$$

we can write

$$dL^2(x,y) = 2L(x,y)du + (\dim M)\,dt.$$

We claim that u is a standard Brownian motion in $\mathbb{R}$. If this is so, then the proof of b) is completed.

To verify the claim we shall apply the Lévy theorem (see, for example, [Doo53]). According to this theorem, it suffices to verify that $u(t)$ and $u^2(t) - t$ are martingales. We observe that

$$L^2 N_m N_n \delta^{mn} = y_i y_j z^i_m z^j_n \delta^{mn} = y_i y_j g^{ij} = L^2,$$

which implies that $N_m N_n \delta^{mn} = 1$. It follows that $u = \int N_n dv^n$ is a square-integrable martingale. By the Itô formula

$$\begin{aligned} du^2 &= 2uN_n dv^n + N_m N_n \delta^{mn} dt \\ &= 2uN_n dv^n + dt, \end{aligned}$$

so $u^2 - t$ is a martingale, which proves the claim. ∎

Remark 3.1. a) An interesting consequence of Theorem 3.12 a) is that if (x, y, z) is a stochastic h-development such that (x, y) starts from a point on the *indicatrix bundle*

$$IM = \{(x, y) \in \overset{\circ}{T}M : L(x, y) = 1\},$$

then (x, y) will P-a.s. remain on IM.

b) Theorem 3.12 b) means that if (x, y, z) is a stochastic hv-development, then $L(x, y)$ exhibits the same behaviour as the radial part $\sqrt{\sum_{n=1}^d (w^n)^2}$ of a standard Brownian motion w in $\mathbb{R}^d$. In particular, this means that P-a.s. $L(x, y)$ can neither escape to ∞ nor to 0 in finite time.

Chapter 4

Volterra-Hamilton Systems of Finsler Type

4.1 Berwald Connections and Berwald Spaces

In this chapter on Volterra-Hamilton theory emphasis will be on the class of Finsler connections known as Berwald connections, which arise locally from the geodesic equations of the Finsler metric function L in that the torsion and curvature tensors are intrinsically given by the equations of the geodesics themselves. This connection is not generally metrical.

Thus, in a local chart we start with the L-geodesic equations

$$\frac{d^2x^i}{ds^2} + 2G^i(x,y) = 0, \quad i = 1, 2, \dots, n, \tag{4.1}$$

where G^i are positive homogeneous of degree two in $y^i = \frac{dx^i}{ds}$, s being the arc-length and $G^i(x,y) = \gamma^i_{jk}(x,y)y^jy^k$, with

$$\gamma^i_{jk}(x,y) = \frac{1}{2}g^{il}\left(\partial_j g_{kl} + \partial_k g_{jl} - \partial_l g_{jk}\right)$$

being the *Christoffel symbols* of the metric tensor g_{ij} for L. The *Berwald connection* is given by

$$\left(N^i_j, F^i_{jk}, V^i_{jk}\right) = \left(G^i_j, G^i_{jk}, 0\right),$$

where $G^i_j = \dot{\partial}_j G^i$ and $G^i_{jk} = \dot{\partial}_k G^i_j = \dot{\partial}_j\dot{\partial}_k G^i$. Consequently, $2G^i = G^i_j y^j = G^i_{jk}y^jy^k$ by Euler's theorem on homogeneous functions. Note that the vertical connection V^i_{jk} is a tensor relative to admissible coordinate changes and that for the Berwald connection it is always vanishing.

This connection has torsion tensors

$$T^i_{jk} = 0, \quad S^i_{jk} = 0, \quad R^i_{jk} = B^i_{0jk}, \quad P^i_{jk} = 0 \tag{4.2}$$

where $B^i_{0jk} := y^h B^i_{hjk}$ with

$$B^i_{hjk} := \delta_k G^i_{hj} + G^m_{hj} G^i_{mk} - (j/k)\,. \tag{4.3}$$

The *deflection tensor* for the Berwald connection vanishes identically in any local chart:

$$D^i_j := y^i{}_{\|j} = \delta_j y^i + G^i_{jk} y^k = -G^i_j + G^i_{jk} y^k \equiv 0. \tag{4.4}$$

The double short bar indicates horizontal covariant differentiation in terms of the Berwald connection.

The corresponding S and P curvatures are

$$S^i_{hjk} = 0, \quad P^i_{hjk} = \dot{\partial}_h G^i_{jk} =: D^i_{hjk}. \tag{4.5}$$

The *Douglas tensor* D^i_{hjk} of the Berwald connection is actually the P curvature of this connection and its vanishing is a necessary and sufficient condition for G^i being a set of n quadratic forms in *any* coordinate system. This is a special circumstance for a Finsler space. Namely, such a space is referred to as being a *Berwald space.* Such spaces play an important role in Volterra-Hamilton theory. The importance of the Douglas tensor is highlighted by

Theorem 4.1. $D^i_{jkl} = 0 \iff C_{ijk|l} = 0 \iff G^i_{jk}$ *is independent of* y, *where the short bar is the Cartan horizontal covariant derivative and* $C_{ijk} = \frac{1}{2}\dot{\partial}_k g_{ij}$ *is the Cartan torsion tensor in pure covariant form* [AIM93].

The condition $C_{ijk|l} = 0$ characterizing Berwald spaces means that C_{ijk} is a parallel tensor in the Cartan sense. In such spaces $g_{ij\|k} = 0$ for Berwald covariant differentiation [AIM93].

Remark 4.1. Coordinates in which Berwald covariant differentiation reduces to ordinary differentiation exist if and only if $D^i_{jkl} = 0$ [Dou28]. These are called *normal coordinates.* There are spaces for which $g_{ij\|k} = 0$ yet there are no normal coordinates. These must be Landsberg spaces which are not Berwald.

A Finsler space is called *locally Minkowski* if there is a covering by local charts in each of which the metric function L is independent of x.

Theorem 4.2. *In a Berwald space* $B^i_{jkl} \equiv 0 \iff B^i_{jk0} \equiv 0 \iff$ *the space is locally Minkowski* [AIM93].

The reader will readily verify

Corollary 4.3. *A Finsler space is locally Minkowski* $\iff D^i_{jkl} \equiv 0$ *and* $B^i_{jkl} \equiv 0$ [AIM93].

In the *two-dimensional case* we can compute C_{ijk} in a simple form by using the *Berwald frame* (l^i, m^i) defined by $l^i := y^i/L(x,y)$, $g_{ij}l^i m^j = 0$, and $g_{ij}m^i m^j = 1$. This defines the frame to within the orientation of m^i. It is easy to see that $g_{ij}l^i l^j = 1$. Furthermore,

$$g_{ij} = l_i l_j + m_i m_j \tag{4.6}$$

for the positive definite case [AIM93]. Here $l_i = g_{ij}l^j$ and $m_i = g_{ij}m^j$. Then, using $C_{ijk}l^k = 0$, from homogeneity we obtain

$$LC_{ijk} = Im_i m_j m_k, \tag{4.7}$$

where $I(x,y)$ is a scalar function homogeneous of degree zero in y called the *principal scalar* of the Finsler space. Obviously,

Corollary 4.4. *A two-dimensional (positive definite) Finsler space is Riemannian* $\iff I(x,y) = 0$.

For the Cartan covariant derivative it is clear that $g_{ij|l} = 0$, $l^i{}_{|j} = 0$, $m^i{}_{|j} = 0$, and $L_{|j} = 0$, which implies

Corollary 4.5. $C_{ijk|l} = 0 \iff I_{|l} = 0$ *for the Cartan covariant derivative.*

Theorem 4.6 (Berwald). $B^i_{jk0} = LKm^i(l_j m_k - l_k m_j)$. *This defines the Gauss-Berwald curvature* $K(x,y)$ *of a (positive definite) two-dimensional Finsler space.*

For a scalar field $S(x,y)$ we obtain Berwald's *h-scalar derivatives* $(S_{,1}, S_{,2})$ and *v-scalar derivatives* $(S_{;1}, S_{;2})$ by setting

$$S_{\|i} = S_{,1}l_i + S_{,2}m_i, \quad S|_i = S_{;1}l_i + S_{;2}m_i, \tag{4.8}$$

where $S_{\|i} = \partial_i S - G^j_i \dot{\partial}_j S$ and $S|_i = \dot{\partial}_i S$. Note that if $S(x,y)$ is positive homogeneous of degree zero in y, then $S_{;1} = 0$. Using this notation on $S = I(x,y)$ in Theorem 4.6, we see that

$$I_{,1,2} - I_{,2,1} = -KI_{;2}. \tag{4.9}$$

From Corollary 4.5 we know that a two-dimensional positive definite Finsler space is Berwald iff $I_{,1} = I_{,2} = 0$. Therefore, $KI_{;2} = 0$, so that

Theorem 4.7 (Berwald). *A positive definite Finsler space of two dimensions is Berwald if and only if it is either locally Minkowski* ($K = 0$) *or* $I = \text{const}$ ($K \neq 0$). *Furthermore, there are three classes of spaces:* (A) $I^2 < 4$, (B) $I^2 = 4$, *and* (C) $I^2 > 4$. *The famous Berwald classification of the above cases is*

$$\begin{aligned} &(\mathrm{A}) \quad L^2 = \left(\beta^2 + \gamma^2\right) e^{2I/\sqrt{4-I^2}\arctan(\gamma/\beta)}, \\ &(\mathrm{B}) \quad L^2 = \beta^2 e^{I\gamma/\beta}, \\ &(\mathrm{C}) \quad L^2 = \beta\gamma\left(\gamma/\beta\right)^{I\sqrt{4-I^2}}, \end{aligned} \tag{4.10}$$

where β *and* γ *are independent one-forms (that is, cross-sections of the cotangent bundle of the Finsler space)* [AIM93]. *See Appendix* B.

We note that (A) has interest in our biological examples below. Antonelli's Berwald metric [Ant91] in ecology is

$$L = e^{(p^2+1)\sigma(x)+p\arctan(y^1/y^2)}\sqrt{(y^1)^2 + (y^2)^2}, \tag{4.11}$$

where σ is a smooth function of x and p is a constant called the *perturbation parameter.* In Sect. 3.1 of [Ant91] $\sigma(x) = \alpha_i x^i$ with constant α_1 and α_2, while in Sect. 3.6 of [Ant91] $\sigma(x) = \alpha_i x^i + \frac{\beta_1}{2}\left(x^1\right)^2 + \beta_{12}x^1x^2 + \frac{\beta_2}{2}\left(x^2\right)^2$ with β_1, β_{12}, β_2 being constants. (C) also have biological applications.

Antonelli showed a relationship between I and p with

Theorem 4.8. *A two-dimensional (positive definite) Finsler space with metric function* (4.11) *is of type* (A) *with* $I = 2p/\sqrt{p^2+1}$. *Furthermore, the Berwald connection coefficients are constants for* (4.11) *iff* $\sigma(x)$ *is linear in* x. *Finally,*

$$K = -\left(p^2+1\right) e^{-\left[2\left(p^2+1\right)\sigma(x)+2p\arctan(y^1/y^2)\right]}\Delta\sigma(x), \tag{4.12}$$

where Δ *is the two-dimensional Euclidean Laplacian.*

Let us prove (4.12). As in [AIM93] (p. 128), the equation for the Landsberg angle θ is

$$L\dot{\partial}_i\theta = m_i \tag{4.13}$$

can be integrated because

$$L\dot{\partial}_j m_i = -\left(l_i - Im_i\right) m_j \tag{4.14}$$

follows from (4.7) and the properties of the Berwald frame (l^i, m^i). Furthermore, The Bianchi identities become quite simply [AIM93]

$$K_{;2} + IK + I_{,1,1} = 0. \tag{4.15}$$

In the case $K \neq 0$ the Berwald space must therefore satisfy (by Theorem 4.7)

$$K_{;2} + IK = 0, \tag{4.16}$$

which upon use of (4.13) becomes

$$\partial_\theta K + IK = 0 \tag{4.17}$$

(see [Run59], p. 256, and [AIM93]). Now, in the case in which $\arctan(y^1/y^2)$ is omitted the metric function (4.11) is Riemannian and it follows that ([Ant85], p. 113)

$$\tilde{K}(x) = -\left(p^2+1\right) e^{-2\sigma(x)\left(p^2+1\right)} \Delta\sigma(x). \tag{4.18}$$

Thus,

$$K(x,y) = \tilde{K}(x) e^{-I\theta}. \tag{4.19}$$

Since from (4.13) we find that

$$\theta = \sqrt{p^2+1}\arctan\left(\frac{y^1}{y^2}\right) \tag{4.20}$$

setting the arbitrary constant of integration equal to zero, we finally obtain (4.12), as desired.

4.2 Volterra-Hamilton Systems and Ecology

Now we describe *n-dimensional Volterra-Hamilton systems* and give some examples. The basic references here are [AIM93] and [AB96]. Let us begin by writing (4.1) in terms of an arbitrary path parameter t:

$$\frac{d^2x^i}{dt^2} + 2G^i_{jk}(x,\dot{x})\frac{dx^j}{dt}\frac{dx^k}{dt} = \frac{\frac{d^2s}{dt^2}}{\frac{ds}{dt}}\frac{dx^i}{dt} \tag{4.21}$$

($\dot{x} = (\dot{x}^1, \ldots, \dot{x}^n)$, $\dot{x}^i = \frac{dx^i}{dt}$). The length s of the trajectory in the Finsler metric is called the *production parameter* and can be interpreted as the *logarithm of total size.* In the case of unlimited or *open growth* $s = Be^{\lambda t}$ solves

$$\frac{d^2s}{dt^2} - \lambda\frac{ds}{dt} = 0, \tag{4.22}$$

allows (4.21) to be written as *Volterra-Hamilton equations*

$$\begin{cases} \dfrac{dx^i}{dt} = N^i, \\ \dfrac{dN^i}{dt} = -G^i_{jk}(x, N)N^jN^k + \lambda N^i. \end{cases} \tag{4.23}$$

Here N^i is interpreted as the *number of individual production units* of kind i and the *Volterra variables*

$$x^i(t) = \int_0^t N^i(\tau)\, d\tau + x^i(0) \tag{4.24}$$

denote the *total accumulation*, up to time t, of the 'product' of kind i.

In general, we replace the first equation by the *Volterra production equation*

$$\frac{dx^i}{dt} = k_{(i)}N^i, \qquad i = 1, \dots, n \tag{4.25}$$

(where the parentheses cancel the summation convention on repeated upper and lower indices), so that $k_i := k_{(i)}$ is the *per capita rate of increase* of x^i. Likewise, the last term of the second equation of (4.23) is generally written as $\gamma^i_j N^j$, where the γ^i_j are constants. The second equation of (4.23) is the *ecological equation.*

The case when the Berwald connection coefficients G^i_{jk} are constants plays an important role in the models of coral communities of n species, [AIM93], [AB96]. Here the coefficients define ecological interactions between these species in a manner completely consistent with the original formulations of V. Volterra, the founder of modern mathematical ecology. The reader is invited to consult any standard text on ecology. We especially recommend G. E. Hutchinson's famous book [Hut78].

The case when the connection coefficients depend only on x^i naturally concerns the Finsler differential geometry of the associated Berwald space playing an important role in *chemical ecology*, which these systems model [AB96]. Especially interesting is the problem of soft coral (*Alcyonacea*) encroachment on hard corals (*Scleractinia*) on reefs such as the Great Barrier

Reef and Pandora Reef. It is the soft coral which has evolved chemicals with which it can kill the reef building hard corals. The variables x^i represent amounts of chemicals in this case and are *allometrically related* to soft coral biomass m (i.e. $x^i = \nu_1^i \log m + \nu_2^i$, $\nu_1^i > 0$). This expresses the fact that such chemicals are uniformly distributed in soft coral tissues [AB96].

The same concepts have been used to model plant/herbivore interactions in the terrestrial environment. In this case the x^i measure so-called defensive compounds which plants have evolved for protection by poisoning herbivores which try to devour them [AB96]. In Sect. 4.7 we discuss stochastic perturbations of the Volterra-Hamilton system (4.23) with metric function (4.11).

4.3 Wagnerian Geometry and Volterra-Hamilton Systems

A contravariant vector A^i is Γ-*parallel* along a smooth curve $C : x = x(t)$ on an n-dimensional manifold with a linear connection $\Gamma(x)$ iff locally

$$\frac{dA^i}{dt} + \Gamma^i_{jk}(x) A^j \frac{dx^k}{dt} = 0. \tag{4.26}$$

The connection can have non-zero torsion tensor $T^i_{jk} := \Gamma^i_{jk} - \Gamma^i_{kj}$ and is defined by the classical *connection transformation law*

$$\frac{\partial \bar{x}^r}{\partial x^j}\frac{\partial \bar{x}^s}{\partial x^k}\bar{\Gamma}^i_{rs}(x) = \frac{\partial \bar{x}^i}{\partial x^r}\Gamma^r_{jk}(x) - \frac{\partial^2 \bar{x}^i}{\partial x^j \partial x^k} \tag{4.27}$$

under a local diffeomorphism from $(x^1, \ldots, x^n)$ to $(\bar{x}^1, \ldots, \bar{x}^n)$. Note that no reference to a metric tensor or function is necessary here.

A vector B^i is Γ-parallel along C to A^i iff

$$\frac{dB^i}{dt} + \Gamma^i_{jk}(x) B^j \frac{dx^k}{dt} = \frac{d \log \varphi}{dt} B^i = f(t) B^i, \tag{4.28}$$

where $B^i = \varphi(t) A^i$ along C and $\varphi(t) = e^{\int_C f(t)\, dt}$.

Let us inquire into the conditions under which parallelism is the same for two linear connections (possibly with torsion) Γ and $\bar{\Gamma}$. It is a classical result that this can happen iff

$$\bar{\Gamma}^i_{jk} = \Gamma^i_{jk} + 2\delta^i_j \psi_k, \tag{4.29}$$

where $\psi_k(x)$ is some covariant vector field [Eis27]. It is not possible for both connections to be torsion-free ($T^i_{jk} = 0$, $\bar{T}^i_{jk} = 0$) in this instance.

Theorem 4.9 (Schouten). *A necessary and sufficient condition that parallelism be the same for two connections Γ^i_{jk} and $\bar{\Gamma}^i_{jk}$, where the first is torsion-free, is that*

$$\bar{T}^i_{jk} = \delta^i_j \psi_k - \delta^i_k \psi_j. \tag{4.30}$$

The connection $\bar{\Gamma}^i_{jk}$ is said to be semi-symmetrically *related to Γ^i_{jk} in this case. Furthermore, the curvature tensors satisfy*

$$\bar{R}^i_{jkl} = R^i_{jkl} + 2\delta^i_j (\partial_k \psi_l - \partial_l \psi_k) \tag{4.31}$$

If $S\bar{\Gamma}^i_{jk}$ denotes the symmetrized connection, *then*

$$S\bar{\Gamma}^i_{jk} = \frac{1}{2}(\bar{\Gamma}^i_{jk} + \bar{\Gamma}^i_{kj}) = \Gamma^i_{jk} + \frac{1}{2}(\delta^i_j \psi_k + \delta^i_k \psi_j) \tag{4.32}$$

The reader will note that $\bar{R}^i_{jkl} = R^i_{jkl}$ iff $\psi_l = \partial_l \psi$ for some smooth scalar function $\psi(x)$.

Now consider the tangent field dx^i/dt along C. $\bar{\Gamma}$-parallelism of these vectors along C means that

$$\frac{d^2x^i}{dt^2} + \bar{\Gamma}^i_{jk} \frac{dx^j}{dt}\frac{dx^k}{dt} = \frac{\frac{d^2s}{dt^2}}{\frac{ds}{dt}} \frac{dx^i}{dt} \tag{4.33}$$

holds according to (4.28) and the fact that

$$\frac{d^2s}{dt^2} - f(t)\frac{ds}{dt} = 0 \tag{4.34}$$

can always be integrated to define a *natural parameter* s for $\bar{\Gamma}$. Namely,

$$s = A + B \int e^{\int_C f(\tau)\, d\tau} dt. \tag{4.35}$$

Choosing this parameter s for t, we obtain

$$\frac{d^2x^i}{ds^2} + \bar{\Gamma}^i_{jk} \frac{dx^j}{ds}\frac{dx^k}{ds} = 0 \tag{4.36}$$

and solutions of these define curves C which are called *autoparallels* of $\bar{\Gamma}$.

Given a semi-symmetric change of Γ into $\bar{\Gamma}$ given by a gradient $\partial_k \psi = \psi_k$, the Γ-autoparallels with natural parameter s will map in one-to-one correspondence with those of $\bar{\Gamma}$ with natural parameter p given by

$$\frac{dp}{ds} = Be^{\int_C \psi(s)\, ds}. \tag{4.37}$$

The autoparallels of $\bar{\Gamma}$ can also be realized as solutions of

$$\frac{d^2x^i}{dp^2} + \left\{\Gamma^i_{jk}(x) + \frac{1}{2}\left(\delta^i_j\psi_k(x) + \delta^i_k\psi_j(x)\right)\right\}\frac{dx^j}{dp}\frac{dx^k}{dp} = 0, \tag{4.38}$$

which are just the autoparallels of the torsion-free connection $S\bar{\Gamma}$.

The trajectories of the Γ-connection are therefore preserved under a semi-symmetric change or, from another point of view, are reparametrized in a way very similar to a *projective change* of connection. We shall use the term *semi-projective transformation* in the semi-symmetric case. It is important to keep in mind the reparametrization perspective for these two concepts for they are fundamental to various models we shall study.

The classical *projective change of connection* [Eis27] is defined by

$$\bar{\Gamma}^i_{jk}(x) = \Gamma^i_{jk}(x) + \delta^i_j\xi_k(x) + \delta^i_k\xi_j(x) \tag{4.39}$$

with

$$\frac{dp}{ds} = Be^{2\int_C \xi_j\, dx^j}. \tag{4.40}$$

Clearly, (4.32) above defines such a change with $\xi_k(x) = \frac{1}{2}\psi_k(x)$.

Now let us consider a Finsler space with fundamental metric tensor g and Finsler connection $(N^i_j, F^i_{jk}, V^i_{jk})$ that satisfies the following four axioms of Hashiguchi:

(H1) It is h- and v-metrical, i.e.

$$g_{ij|k} = 0 \quad \text{and} \quad g_{ij}|_k = 0,$$

(H2) The deflection tensor D vanishes identically, i.e.

$$D^i_k = y^jF^i_{jk} - N^i_k = 0,$$

(H3) It is semi-symmetric, i.e.

$$T^i_{jk} = F^i_{jk} - F^i_{kj} = \delta^i_j\sigma_k - \delta^i_k\sigma_j$$

for some specified covariant Finsler vector field $\sigma_j(x, y)$,

(H4) The vertical torsion tensor vanishes identically, i.e.

$$S^i_{jk} = V^i_{jk} - V^i_{kj} = 0.$$

The connection is uniquely determined [Has75] and is called the *Wagner connection* relative to $\sigma_i(x, y)$. Note that $V^i_{jk} = C^i_{jk}$, the Cartan tensor. If $\sigma_j = \partial_j\sigma$ for some smooth scalar function and if F^i_{jk} is independent of y, the space is called a *Wagner space relative to* σ (sometimes, σ-*Wagner space*). $F^i_{jk}(x)$ is a linear connection.

Note that any locally Minkowski space is a Wagner space relative to the zero function. More generally, a Berwald space is a Wagner space relative to the zero function.

Theorem 4.10 (Hashiguchi). *If* $(N^i_j, F^i_{jk}(x), C^i_{jk})$ *is the Wagner connection of a Wagner space with metric function* $L(x, y)$ *and* $\bar{L} = e^{\sigma(x)}L$ *is a conformally related metric, it is also a Wagner space equipped with connection* $(\bar{N}^i_j, \bar{F}^i_{jk}(x), \bar{C}^i_{jk}) = (N^i_j + y^i\partial_j\sigma, F^i_{jk} + \delta^i_j\partial_k\sigma, C^i_{jk})$ *with* $\bar{T}^i_{jk} = T^i_{jk} + \delta^i_j\partial_k\sigma - \delta^i_k\partial_j\sigma$.

Corollary 4.11. *If* $L(x, y)$ *is a locally Minkowski metric function, then any conformally related metric* $\bar{L} = e^{\sigma(x)}L$ *is a* σ-*Wagner space with Wagner curvature zero.*

Proof. Any Wagner space arising from another by a conformal change of the metric function as above induces a semi-projective transformation of the geodesics via the gradient $\partial_j\sigma(x)$. Consequently, $\bar{R}^i_{jkl} = R^i_{jkl}$ by (4.31). Since the Wagner connection for a locally Minkowski space can be expressed as $(0, 0, C^i_{jk})$, it follows that $R^i_{jkl} \equiv 0$ and the corollary is proved. ∎

Remark 4.2. The converse of Corollary 4.11 holds: A Wagner space with $\sigma_i(x) = \partial_i\sigma(x)$ for which the Wagner curvature vanishes is necessarily conformally locally Minkowski with $\bar{L} = e^{\sigma(x)}L$ [Has75].

We now turn to examples of 2-dimensional Wagner spaces. The basic reference is [Mat82].

Theorem 4.12 (Wagner). *Let* I *and* θ *denote the principal scalar and Landsberg angle of a 2-dimensional Finsler space and suppose that* $\partial_\theta I \neq 0$. *This space is a Wagner space iff* $\partial_\theta I$ *is a function of* I. *Furthermore, the vector field* σ_i *is given by*

$$\sigma_1 = -\frac{I_{,2} + II_{,1}}{I_{;2}}, \quad \sigma_2 = \frac{I_{,1}}{I_{;2}}. \tag{4.41}$$

Remark 4.3. It is well known that I and θ are conformal invariants [IH77].

For the mth-root metric function [AS91]

$$L(x,y) = \left[(y^1)^m + (y^2)^m\right]^{1/m} \tag{4.42}$$

we obtain

$$I = \frac{m-2}{2\sqrt{m-1}} \frac{1-z^m}{\sqrt{z^m}}, \tag{4.43}$$

where $z = y^2/y^1$. Note that I may have the opposite sign if the Berwald frame vector m^i has reversed orientation. However, I^2 is always uniquely defined. Since the Landsberg angle is [AS91]

$$\theta = \frac{2\sqrt{m-1}}{m} \arctan \sqrt{z^m}, \tag{4.44}$$

we see that

$$\partial_\theta I = -\frac{m(m-2)}{4(m-1)} \cdot \frac{(1+z^m)^2}{z^m}.$$

For $m \neq 2$ we obtain, therefore,

$$\partial_\theta I = -\left[\frac{m(m-2)}{m-1} + \frac{m}{m-2} I^2\right]. \tag{4.45}$$

If we set $m = 3$, we obtain

$$\partial_\theta I = -\frac{3}{2} - 3I^2, \tag{4.46}$$

which was originally obtained by Wagner for any *cubic metric function*

$$L(x,y) = \left[a_{ijk}(x) y^i y^j y^k\right]^{1/3}. \tag{4.47}$$

Of course it is only the case when m is even that yields positive definite metric tensors necessary for the stochastic theories developed herein.

For general 2-dimensional Finsler spaces we have the following result [AZ96].

Theorem 4.13 (Antonelli, Zastawniak). *For the Finsler metric function $\bar{L} = e^{\sigma(x)} L$, where L is locally Minkowskian, the Berwald-Gauss curvature is expressed as*

$$\bar{K} = -\left(\bar{g}^{ij}\sigma_{ij} + I\bar{m}^i \bar{l}^j \sigma_{ij} - (\partial_\theta I)\bar{m}^i \bar{m}^j \sigma_i \sigma_j\right), \tag{4.48}$$

where $\bar{g}^{ij} = e^{-2\sigma(x)} g^{ij}$ and $\bar{m}^i = e^{-\sigma(x)} m^i$, (l^i, m^i) being the Berwald frame for L and $\sigma_{ij}(x) := \partial_i \partial_j \sigma(x)$.

The reader may compute (4.12) from (4.48).

For $\bar{L} = e^{\sigma(x)}L$ with $\sigma(x) = \alpha_i x^i$ and L as in (4.42) we have [AS91]

$$\bar{K} = \frac{m(m-2)}{4(m-1)^2}\frac{\alpha_1^2\left(z^{m-1} - \alpha_1/\alpha_2\right)^2(z^m+1)^2}{z^{2m-1}}\frac{y^1y^2}{\bar{L}^2}. \tag{4.49}$$

If we start with a locally Minkowski metric L and perform a projective change of connection via $\phi(x)$, we arrive at reparametrized straight lines

$$\frac{d^2x^i}{dp^2} + \left(\delta_j^i\phi_k + \delta_k^i\phi_j\right)\frac{dx^j}{dp}\frac{dx^k}{dp} = 0, \tag{4.50}$$

where the new 'tick' of time or production measure is

$$p = A + B\int e^{2\int_\gamma \phi(\tau)\,d\tau}\,dt$$

along a trajectory γ.

This serves as a model of *heterochrony*, time-sequencing change in growth and development for n-dimensional Gompertz growth

$$\frac{d^2x^i}{dt^2} + \lambda\frac{dx^i}{dt} = 0, \tag{4.51}$$

called *Laird's law* [AB96]. First, we pass to the total production parameter s so that on γ

$$ds = L(x, dx) \quad \text{or} \quad L(x, dx/ds) = 1$$

and (4.51) becomes

$$\frac{d^2x^i}{ds^2} = 0. \tag{4.52}$$

Then we arrive at (4.50) and finally pass back to real time t to obtain the Volterra-Hamilton system

$$\left\{\begin{aligned} \frac{dx^i}{dt} &= N^i, \\ \frac{dN^i}{dt} &= -\left(\delta_j^i\phi_k + \delta_k^i\phi_j\right)N^jN^k + \bar{\lambda}N^i, \end{aligned}\right. \tag{4.53}$$

where $\bar{\lambda}$ is some constant.

A heterochronic change can be 'reversible,' as the above projective change procedure can be reversed, but usually such time-sequencing changes occur because of specific environmental influences. Thus, we need to have some description of these dynamic external influences. As a first attempt we could try to add $\phi^i = g^{ij}\phi_j = g^{ij}\partial_j\phi$ to (4.50). But because we must require the heterochronically transformed system to be derivable from a Lagrangian, in fact a Finslerian metric, we will set $dp = \bar{L}(x, dx)$ and try to determine $\bar{L}$. Since $\bar{L}(x, dx/dp) = 1$ along trajectories now, we observe that $\phi^i = L^2\phi^i = L^2 g^{ij}\partial_j\phi$ and that (4.50) is therefore equivalent to

$$\frac{d^2x^i}{dp^2} + \left(\delta^i_j\phi_k + \delta^i_k\phi_j - g_{jk}\phi^i\right)\frac{dx^j}{dp}\frac{dx^k}{dp} = 0 \tag{4.54}$$

with

$$\bar{L}^2 = e^{2\phi} g_{jk}\frac{dx^j}{dp}\frac{dx^k}{dp} = e^{2\phi}L^2\left(\frac{dp}{ds}\right)^{-2} = \bar{L}^2 = 1.$$

In other words, (4.54) are the geodesic equations of the Finsler metric $e^{\phi(x)}L = \bar{L}$. Thus, a *heterochronic transformation* will be a passage from (4.51) to

$$\frac{d^2x^i}{dp^2} + \left(\delta^i_j\phi_k + \delta^i_k\phi_j\right)\frac{dx^j}{dp}\frac{dx^k}{dp} = \phi^i. \tag{4.55}$$

It is known that the external environmental influences cause the solution of (4.51) to become curved. But we are not able to express this in the obvious way, i.e. by using normal coordinates for $\bar{L} = e^{\phi}L$ at an arbitrary point. However, if we use Wagnerian geometry, we can rewrite the geodesics of $\bar{L}$ as

$$\begin{aligned}
\frac{d^2x^i}{dp^2} + \left(\delta^i_j\phi_k\right)\frac{dx^j}{dp}\frac{dx^k}{dp} &= \bar{L}^2\bar{\phi}^i - \left(\delta^i_k\phi_j\right)\frac{dx^j}{dp}\frac{dx^k}{dp} \\
&= C^i\left(x, \frac{dx}{dp}\right),
\end{aligned} \tag{4.56}$$

where

$$\bar{\phi}^i\bar{L}^2 = e^{2\phi}L^2\bar{g}^{ij}\phi_j = L^2 g^{ij}\phi_j = L^2\phi^i.$$

Now the left-hand side can be written in terms of $S\Gamma$, the symmetrized Wagner connection, as

$$\frac{d^2x^i}{dp^2} + (S\Gamma)^i_{jk}\frac{dx^j}{dp}\frac{dx^k}{dp} = C^i\left(x, \frac{dx}{dp}\right). \tag{4.57}$$

At an arbitrary point we use normal coordinates $\tilde{x}^i$ for $S\Gamma$ to obtain just

$$\frac{d^2\tilde{x}^i}{dp^2} = C^i \tag{4.58}$$

for the geodesics through this point. Thus, (4.57) expresses a complete concept of heterochrony, one that embodies a semi-projective (time-sequencing) change of parameter along the Gompertz trajectories of (4.51) together with external environmental causative influence C^i. The reader can verify that C^i is orthogonal (relative to g_{ij} or $\bar{g}_{ij}$) to the solutions of (4.57). Notice that we have not claimed that ϕ^i is orthogonal to these geodesics. This is one reason why normal coordinates for the symmetric linear connection $\delta^i_j\psi_k + \delta^i_k\psi_j$ do not give the correct description of how environmental influences cause the original straight lines to become curved upon heterochronic transformation.

Remark 4.4. Normal coordinates exist for the Wagner connection of Wagner space in the sense that $S\Gamma$ in (4.57) can be viewed as a torsion-free linear connection. Moreover, for the special case of the mth-root metric L defined by (4.42) with

$$\bar{L} = e^{\sigma(x)}L, \quad \sigma(x) = \alpha_i x^i, \quad \alpha_1, \alpha_2 > 0, \tag{4.59}$$

the x^i being log biomasses, one can exhibit a consistent interpretation of these normal coordinates as multiples of true logarithms of biomasses. The reason is that the heterochronic transformation of the straight lines of the L-space to the geodesics of $\bar{L}$ as in (4.57) actually preserves a true straight line out of the origin of the L-space. This is because $C^i = 0$ has exactly one solution, the ratio $y^1/y^2 = (\alpha_1/\alpha_2)^{1/(m-1)}$, so $\tilde{x}^1 = (\alpha_1/\alpha_2)^{1/(m-1)}\tilde{x}^2$, and this is the image of $x^1 = (\alpha_1/\alpha_2)^{1/(m-1)}x^2$ in L-space, so $\tilde{x}^1 = ax^1$ and $\tilde{x}^2 = bx^2$ for some constants a, b. This gives the consistency of interpretation.

The geodesics of (4.59) have interesting ecological meaning because the interaction coefficients are not constant ($n \geq 3$) but do not depend on x^i. Such spaces are called y-Berwald spaces [Ant93].

4.4 Random Perturbations of Finslerian Volterra-Hamilton Systems

We consider a general Volterra-Hamilton system of Finsler type

$$\begin{cases} \dfrac{dx^i}{ds} = y^i, \\ \dfrac{dy^i}{ds} = -F^i_{jk}(x,y)y^j y^k, \end{cases} \tag{4.60}$$

using the total production parameter s. Here F^i_{jk} are the horizontal connection coefficients of a Finsler connection $\left(F^i_{jk}, N^i_j, C^i_{jk}\right)$ that is h- and v-metrical relative to a Finsler metric function $L(x,y)$ with the corresponding metric tensor $g_{ij}(x,y)$. As explained earlier in this chapter, the variables y^i can be interpreted as population numbers and the x^i as the (log-)biomasses.

In the most general situation, random environmental and/or developmental perturbations can occur in both the production and population equations. This will be expressed by adding noise terms to the right-hand sides of both equations in (4.60). In our model we take two independent standard (Euclidean) Brownian motions $v^i(s)$ and $w^i(s)$ in $\mathbb{R}^n$ perturbing the biomasses $x^i(s)$ and population numbers $y^i(s)$, respectively. We assume that this external noise $v^i(s), w^i(s)$ is not affected by the behaviour of the Volterra-Hamilton system. However, the response $dx^i(s), dy^i(s)$ to the perturbations will, in general, depend not only on the magnitude and direction of infinitesimal perturbations $dv^i(s), dw^i(s)$, but also on state $x^i(s), y^i(s)$ of the system.

But what are the responses $dx^i(s), dy^i(s)$ as functions of $dv^i(s), dw^i(s)$? In the present section we address this question by establishing a *general rule for noise addition* in Volterra-Hamilton systems of Finsler type.

The metric plays a key role in expressing the responses in terms of perturbations. Loosely speaking,

> *the infinitesimal distance by which the state (x,y) of the Volterra-Hamilton will be displaced must be proportional to the magnitude of the perturbation dv^i, dw^i*

Since $(x,y) \in TM$, a suitable metric on the tangent bundle TM is necessary to measure this distance. A natural choice is the diagonal lift $G(x,y)$ of the Finsler metric tensor $g(x,y)$ as described in Chap. 3. Thus, given the perturbations dv^i, dw^i with $dv^i dv^j = dw^i dw^j = \delta^i_j ds$ and $dv^i dw^j = 0$, the rule for noise addition can be expressed by the condition that

> *the covariance matrix of x^i, dy^i must be equal to the metric of the reciprocal tensor to $G(x,y)$ in the natural frame $\partial_i, \dot{\partial}_j$ times ds, that is,*

$$\begin{bmatrix} dx^i dx^j & dx^i dy^j \\ dy^i dx^j & dy^i dy^j \end{bmatrix} = \begin{bmatrix} g^{ij} & -g^{ik} N^j_k \\ -g^{jk} N^i_k & g^{ij} + g^{kl} N^i_k N^j_l \end{bmatrix} ds \qquad (4.61)$$

This can be achieved with the aid of *hv*-isometrical rolling (see Chap. 3). To begin with, we consider a solution $x(s), y(s)$ to the deterministic system (4.60). Then *hv*-rolling along this solution defines a smooth curve $\mu(s), \nu(s)$ in $\mathbb{R}^{2n}$ given by the system of differential equations

$$\begin{cases} \dfrac{dx^i}{ds} = y^i + z^i_j \dfrac{d\mu^j}{ds}, \\ \dfrac{\delta y^i}{ds} = z^i_j \dfrac{d\nu^j}{ds}, \\ \dfrac{dz^i}{ds} = -F^i_{jk}(x,y) z^l_j \dfrac{dx^k}{ds} - C^i_{jk}(x,y) z^l_j \dfrac{\delta y^k}{ds}, \end{cases}$$

where

$$\frac{\delta y^i}{ds} = \frac{dy^i}{ds} + N^i_j(x,y)\frac{dx^j}{ds}.$$

Next we perturb $\mu(s), \nu(s)$ by adding $v(s), w(s)$ and use *hv*-isometrical rolling (now the stochastic version of it) to transfer the resulting diffusion $\mu(s) + v(s), \nu(s) + w(s)$ back to the manifold. Retaining the same notation $x(s), y(s)$ for the process on the manifold, we obtain the system of stochastic differential equations

$$\begin{cases} dx^i = y^i ds + z^i_j \circ dv^j, \\ \delta y^i = z^i_j \circ dw^j, \\ dz^i = -F^i_{jk}(x,y) z^l_j \circ dx^k - C^i_{jk}(x,y) z^l_j \circ \delta y^k, \end{cases} \tag{4.62}$$

where

$$\delta y^i = dy^i + N^i_j(x,y) \circ dx^j. \tag{4.63}$$

It is now a matter of straightforward computation to verify that dx^i, dy^i satisfy (4.61). This means that (4.62) defines a diffusion consistent with our rule for noise addition. The use of Stratonovich stochastic calculus ensures that the covariance of the Volterra-Hamilton system is carried over to the system with noise.

4.5 Random Perturbations of Riemannian Volterra-Hamilton Systems

In this section we specialize to Riemannian Volterra-Hamilton systems of the form

$$\begin{cases} \dfrac{dx^i}{ds} = y^i, \\ \dfrac{dy^i}{ds} = -\Gamma^i_{jk}(x)y^iy^j, \end{cases} \tag{4.64}$$

where Γ^i_{jk} are the coefficients of a metrical connection under a Riemannian metric tensor $g_{ij}(x)$. This is a much simpler setting than general Volterra-Hamilton systems of Finsler type. To explain the role of our rule for noise addition we shall study the simpler Riemannian case first, before proceeding to Finslerian examples in the following sections.

Because in the Riemannian case the metric tensor $g_{ij}(x)$ is independent of y, the fibres TM_x of the tangent bundle have the structure of a *flat* Euclidean space. Therefore, a natural choice of the metric on the tangent bundle TM is

$$[G_{AB}] = \begin{bmatrix} g_{ij} & 0 \\ 0 & \delta_{\bar{\imath}\bar{\jmath}} \end{bmatrix}, \tag{4.65}$$

where g_{ij} is the given Riemannian metric on M, δ_{ij} is the Kronecker symbol, $A, B = 1, \dots, 2n$, $i, j = 1, \dots, n$, and $\bar{\imath}, \bar{\jmath} = n+1, \dots, 2n$. Clearly, G_{AB} defines a Riemannian metric on TM.

We can now apply the general rule for noise addition by requiring that the covariance matrix of dx^i, dy^i should be equal to the reciprocal tensor G^{AB} to G_{AB} times ds, that is,

$$\begin{bmatrix} dx^idx^j & dx^idy^j \\ dy^idx^j & dy^idy^j \end{bmatrix} = \begin{bmatrix} g^{ij} & 0 \\ 0 & \delta^{ij} \end{bmatrix} ds \tag{4.66}$$

This is a modification of the general Finslerian case (4.61), for the much simpler metric (4.66) is used in place of the diagonal lift metric.

Condition (4.66) is fulfilled by a diffusion satisfying the following system of stochastic differential equations:

$$\begin{cases} dx^i = y^ids + z^i_j \circ dv^j, \\ dy^i = -\Gamma^i_{jk}(x)y^jy^kds + dw^i, \\ dz^i_j = -\Gamma^i_{lk}(x)z^l_j \circ dx^k, \end{cases} \tag{4.67}$$

Here, as before, $v^i(s)$ and $w^i(s)$ are independent standard Brownian motions in $\mathbf{R}^n$ and $z^i_j(s)$ is an auxiliary orthonormal frame process.

The resulting Markov diffusion $x(s), y(s)$ on TM has generator

$$\mathcal{G} = \frac{1}{2} g^{ij}\left(\partial_i\partial_j - \Gamma^k_{ij}\partial_k\right) + \frac{1}{2}\delta^{ij}\dot{\partial}^i\dot{\partial}^j + y^i\partial_i - \Gamma^i_{jk}(x) y^j y^k \dot{\partial}_i. \tag{4.68}$$

The probability density $\rho(s,x,y)$ of $x(s), y(s)$ satisfies the initial-boundary value problem

$$\begin{cases} \dfrac{\partial \rho}{\partial s} = \mathcal{G}^*\rho, \\ \lim\limits_{s\searrow 0} \rho(s,x,y) = \rho_0(x,y), \\ \rho(s,x,y)|_{\partial D} = 0, \end{cases} \tag{4.69}$$

where the initial density $\rho_0(x,y)$ is supported on $D = \{(x,y) \in TM : y^i > 0,\ i = 1,\dots,n\}$ and $\mathcal{G}^*$ is the formally adjoint operator to $\mathcal{G}$ relative to the metric (4.65). Obviously, D is the domain in which the biological interpretation of the y^i as population sizes is applicable.

Example 4.1 (conformally flat Riemannian metric). A metric $g_{ij}(x)$ on $\mathbf{R}^n$ is *conformally flat* if there is a smooth function $\phi(x)$ on $\mathbf{R}^n$ such that

$$g_{ij}(x) = e^{2\phi(x)}\delta_{ij}. \tag{4.70}$$

(Here we consider a fixed orthonormal coordinate system in $\mathbf{R}^n$.) The Levi-Cività connection coefficients of g_{ij} are easily computed to be

$$\begin{array}{ll} \gamma^i_{ii} = -\gamma^i_{jj} = \partial_i\phi & \text{if } i \neq j, \\ \gamma^i_{ij} = \gamma^i_{ji} = \partial_j\phi & \text{if } i \neq j, \\ \gamma^i_{jk} = 0 & \text{otherwise.} \end{array}$$

The corresponding Volterra-Hamilton system describes the growth of a colony of n symbionts, where each of the n subpopulations interacts with at least one of the other subpopulations. The case when the connection coefficients (i.e. the coefficients in the Volterra-Hamilton system) are constant, that is, when $\phi(x) = \alpha_i x^i$ is linear has been studied extensively in [AIM93], [AB96]. However, here we consider the general case when $\phi(x)$ is an arbitrary function with non-vanishing $\partial_i\phi$ for all $i = 1,\dots,n$.

Setting $\Gamma^i_{jk} = \gamma^i_{jk}$ in (4.67), we obtain the perturbed process $x(s), y(s)$, whose generator can be evaluated from (4.68) to be

$$\mathcal{G}f = \frac{1}{2}\Delta_G f + y^i\partial_i f - \partial_j\phi\left(2y^i y^j - \delta^{ij}|y|^2\right)\dot{\partial}_i f.$$

We shall compute the adjoint operator $\mathcal{G}^*$ appearing in the initial-boundary value problem (4.69) in terms of the scalar curvature $\mathcal{R}$ of g_{ij} (which is identical to the Riemannian scalar curvature of the metric G_{AB} on TM defined by (4.66) because the fibres TM_x are flat). To this end we recall that

$$\mathcal{R} = g^{ij}\mathcal{R}_{ij},$$

$\mathcal{R}_{ij}$ being the Ricci curvature tensor given by

$$\mathcal{R}_{ij} = \partial_k\Gamma^k_{ij} - \Gamma^r_{ik}\Gamma^k_{rj} + \Gamma^k_{ij}\partial_k\left(\ln\sqrt{g}\right) - \partial_i\partial_j\left(\ln\sqrt{g}\right),$$

where $g = \det[g_{ij}] = e^{2n\phi}$. Thus we find that

$$\mathcal{R} = -(n-1)e^{-2\phi}\delta^{ij}\left[2\partial_i\partial_j\phi + (n-2)\partial_i\phi\partial_j\phi\right].$$

Now, a straightforward if somewhat tedious calculation shows that the adjoint operator to $\mathcal{G}$ is

$$\mathcal{G}^*f = \frac{1}{2}\Delta_G f + A^i\partial_i f + B^i\dot{\partial}_i f + Vf,$$

where

$$\begin{aligned} A^i &= -y^i, \quad B^j = \partial_i\phi\left(2y^iy^j - \delta^{ij}|y|^2\right), \\ \frac{V}{n} &= \left(\frac{\mathcal{R}}{2(n-1)} + \frac{n+2}{2}|\nabla_g\phi|^2\right) + 2\partial_i\phi y^i. \end{aligned} \tag{4.71}$$

The forward initial-boundary value problem (4.69) for $\rho(s,x,y)$ can now be solved with the aid of the Feynman-Kac formula. To this end we introduce an auxiliary diffusion (X,Y) on TM defined by the following system of stochastic defferential equations:

$$\left\{\begin{aligned} dX^i &= A^i ds + Z^i_j \circ dv^j, \\ dY^i &= B^i ds + dw^i, \\ dZ^i &= -\Gamma^i_{jk}(X)Z^l_j \circ dX^k, \end{aligned}\right. , \tag{4.72}$$

where, as usual, $Z^i_j(s)$ s an auxiliary orthonormal frame process. To ensure the existence of solutions of (4.72) without explosions one can employ the bump function technique, assuming that ϕ is compactly supported with $\operatorname{supp}\phi$ large enough to contain the whole region of biological interest with

$\partial_i\phi$ non-vanishing anywhere in that region for any $i = 1, \ldots, n$. By the Feynman-Kac formula, the solution to (4.69) can be represented as

$$\rho(s,x,y) = E_{x,y}\left\{\chi_{\{\sigma>s\}}\rho_0(X(s),Y(s))\exp\left[\int_0^s V(X(r),Y(r))\,dr\right]\right\}, \tag{4.73}$$

where $\sigma = \inf\{s \geq 0 : (X(s), Y(s)) \in \partial D\}$ is the first time of hitting the boundary of D and $E_{x,y}$ is the conditional expectation given that $X(0) = x$ and $Y(0) = y$.

The Feynman-Kac representation of $\rho(s, x, y)$ can be used, in particular, to study the effect of the scalar curvature $\mathcal{R}$ on the behaviour of the perturbed Volterra-Hamilton system. In the deterministic case the solutions to the Volterra-Hamilton system are stable if and only if the Riemannian scalar curvature $\mathcal{R}$ is positive everywhere (after Jacobi). In the case of random perturbations when formula (4.73) applies, it follows by (4.71) that, for s small enough, $\rho(s, x, y)$ will increase locally as $\mathcal{R}$ increases towards positive values, all other thing being equal, while $\mathcal{R}$ decreasing towards large negative values will have the opposite effect. This leads to a picture consistent with the stability results for deterministic systems.

Example 4.2. (stochastic perturbations giving rise to the Wagner connection). It is important to point out that the rule for noise addition leaves some freedom in defining the explicit form of the perturbation terms. Even though isometrical rolling as described by (4.67) provides a general prescription for noise addition, in some cases simpler noise terms can be found which are also consistent with (4.65) and biologically relevant. In particular, in the case of the conformally flat metric (4.70) the following system of stochastic defferential equations can be used instead of (4.67):

$$\begin{cases} dx^i = y^i ds + e^{-\phi} \circ dv^i, \\ dy^i = -\Gamma^i_{jk}(x) y^j y^k ds + dw^i. \end{cases} \tag{4.74}$$

Again, one can readily verify using Itô's formula that the covariance matrix of dx^i, dy^i satisfies (4.66), as required.

The perturbation term $du^i = e^{-\phi} \circ dv^i$ is closely related to the Wager connection with coefficients $F^i_{jk} = \partial_j\phi\delta^i_k$ (cf. Sect. 4.3). Namely, for any smooth function $f(x)$,

$$Edf(u(s)) = \frac{1}{2} g^{ij}\left(\partial_i\partial_j f + F^k_{ij}\partial_k f\right) ds.$$

If $u(s)$ taken on its own were a Markov process (in particular, if the term $y^d s$ were absent in (4.74), then $\frac{1}{2}g^{ij}\left(\partial_i\partial_j f + F^k_{ij}\partial_k f\right)$ would be its generator. This observation has important consequences, for the system (4.74) can be written in terms of the Wagner connection as

$$\begin{cases} dx^i = y^i ds + e^{-\phi} \circ dv^i, \\ dy^i + F^i_{jk}(x) y^j y^k ds = C^i(x,y) + dw^i, \end{cases} \tag{4.75}$$

where C^i is given by (4.57). The second equation in (4.75) can clearly be understood as the geodesic equation (4.56) written in terms of Wagnerian geometry and perturbed by noise.

4.6 Noise in Conformally Minkowski Systems

The metric function $\bar{L}$ of a conformally Minkowski space is defined by

$$\bar{L}(x,y) = e^{\phi(x)} L(y), \quad x \in \mathbb{R}^n, \quad y \in \mathbb{R},$$

where $L(x,y) = L(y)$ is homogeneous of degree one in y, i.e. it is the metric function of an n dimensional Minkowski space. In fact, it is enough to require $L(x,y)$ to be locally Minkowski, i.e. $L(x,y) = L(y)$ in some system of coordinates x. Clearly, $\bar{L}(x,y)$ defines the structure of a Finsler manifold on $\mathbb{R}^n$ called a *conformally Minkowski space.*

The corresponding metric tensors

$$\bar{g}_{ij} = \frac{1}{2}\dot{\partial}_i\dot{\partial}_j \bar{L}^2 \quad \text{and} \quad g_{ij} = \frac{1}{2}\dot{\partial}_i\dot{\partial}_j L^2$$

are related by

$$\bar{g}_{ij}(x,y) = e^{2\phi(x)} g_{ij}(y),$$

in analogy to the conformally flat case (4.70), which makes it possible to construct random perturbations in much the same way as in the examples in Sect. 4.5. In particular, we refer to Example 4.2, which establishes a relationship between random perturbations and the Wagner connection.

The Finsler space with metric $\bar{L}$ has the structure of a Wagner space with the natural Wagner connection $\left(\bar{F}^i_{jk}, \bar{N}^i_j, C^i_{jk}\right)$, where (see Sect. 4.3)

$$\bar{F}^i_{jk} = \partial_k \phi \delta^i_j, \quad \bar{N}^i_j = \partial_j \phi y^i, \quad C^i_{jk} = \frac{1}{2}\bar{g}^{il}\dot{\partial}_j \bar{g}_{jk} = \frac{1}{2} g^{il}\dot{\partial}_j g_{jk}.$$

(The vertical connection coefficients C^i_{jk} are written without a bar because they are conformal invariants.) The connection coefficients F^i_{jk} and N^i_j of the Minkowski metric $L(y)$ are of course identically equal to zero.

The geodesic equations can be written in terms of the Wagner connection as (see (4.56)

$$\begin{cases} \dfrac{dx^i}{ds} = y^i, \\ \dfrac{\bar{\delta} y^i}{ds} = C^i, \end{cases} \tag{4.76}$$

where

$$\frac{\bar{\delta} y^i}{ds} = \frac{dy^i}{ds} + \bar{N}^i_j \frac{dx^j}{ds}$$

and

$$C^i = g^{ij} g_{km} y^k y^m \partial_j \phi - y^i y^j \partial_j \phi. \tag{4.77}$$

Because the fibres of the tangent bundle are no longer Euclidean, but have the structure of a Riemannian manifold with metric function $L(y)$, metric tensor $g_{ij}(y)$ and connection coefficients $C^i_{jk}(y)$, it is necessary to modify the system (4.75) of stochastic defferential equations by introducing an auxiliary frame z^i_j orthonormal under g_{ij}:

$$\begin{cases} dx^i = y^i ds + e^{-\phi} z^i_j \circ dv^j, \\ \bar{\delta} y^i = C^i ds + e^{-\phi} z^i_j \circ dw^j, \\ dz^i_j = -C^i_{kl} z^l_j \circ dy^k. \end{cases} \tag{4.78}$$

This system combines the idea of isometrical rolling of a fibre regarded as a Riemannian manifold with metric tensor $g_{ij}(y)$ and Levi-Civitá coefficients $C^i_{jk}(y)$ with a stochastic perturbation of the form put forward in Example 4.2.

One can verify that

$$\begin{aligned} dx^i dx^j &= e^{-2\phi} z^i_k z^j_l \delta^{kl} ds = e^{-2\phi} g^{ij} ds = \bar{g}^{ij} ds, \\ dx^i dy^j &= dx^i (\bar{\delta} y^j - \bar{N}^j_k \circ dx^k) = -\bar{N}^i_k e^{-2\phi} g^{kj} ds = -\bar{N}^i_k \bar{g}^{kj} ds, \\ dy^i dy^j &= (\bar{\delta} y^i - \bar{N}^i_k \circ dx^k)(\bar{\delta} y^j - \bar{N}^j_k \circ dx^k) = e^{-2\phi}(g^{ij} + g^{kl} \bar{N}^i_k \bar{N}^j_l) ds \\ &= (\bar{g}^{ij} + \bar{g}^{kl} \bar{N}^i_k \bar{N}^j_l) ds, \end{aligned}$$

which means that the rule of noise addition (4.61) is satisfied for the diagonal lift $\bar{G}(x, y)$ of the of the metric tensor $\bar{g}(x, y)$.

The generator of $x(s), y(s)$ defined by (4.78) is easily seen to be

$$\mathcal{G} = \frac{1}{2}\bar{g}^{ij}\left(\bar{\delta}_i\bar{\delta}_j - \bar{F}^k_{ij}\bar{\delta}_k\right) + \frac{1}{2}\bar{g}^{ij}\left(\dot{\partial}_i\dot{\partial}_j - C^k_{ij}\dot{\partial}_k\right) + y^i\bar{\delta}_i + C^i\dot{\partial}_i. \tag{4.79}$$

Remark 4.5. An alternative way of introducing random perturbations in (4.76) is to use the hv-isometrical rolling with respect to the Wagner connection $\left(\bar{F}^i_{jk}, \bar{N}^i_j, C^i_{jk}\right)$:

$$\left\{\begin{array}{l} dx^i = y^i ds + z^i_j \circ dv^j, \\ \bar{\delta}y^i = C^i ds + z^i_j \circ dw^j, \\ dz^i_j = -\bar{F}^i_{kl} z^l_j \circ dx^k - C^i_{kl} z^l_j \circ \bar{\delta}y^k. \end{array}\right. \tag{4.80}$$

Here the auxiliary frames z^i_j are orthonormal in the metric $\bar{g}_{ij}$ (rather than g_{ij}, as in (4.78). The Markov diffusion $x(s), y(s)$ defined by (4.80) is equivalent to that defined by (4.78) in the sense that it has the same generator (4.79).

4.7 Canalization of Growth and Development with Noise

Consider a system governed by the Finsler Lagrangian (4.11), i.e.

$$L^2(x,y) = \left[(y^1)^2 + (y^2)^2\right] e^{2(p^2+1)\alpha(x)+2p\arctan(y^1/y^2)}, \tag{4.81}$$

where $x = (x^1, x^2)$ and $y = (y^1, y^2)$. The corresponding Euler-Lagrange equations

$$\begin{aligned} &\frac{d^2x^1}{ds^2} + 2(\alpha_2 + \alpha_1 p)\frac{dx^1}{ds}\frac{dx^2}{ds} + (\alpha_1 - \alpha_2 p)\left[\left(\frac{dx^1}{ds}\right)^2 - \left(\frac{dx^2}{ds}\right)^2\right] = 0, \\ &\frac{d^2x^2}{ds^2} + 2(\alpha_1 - \alpha_2 p)\frac{dx^1}{ds}\frac{dx^2}{ds} + (\alpha_2 - \alpha_1 p)\left[\left(\frac{dx^2}{ds}\right)^2 - \left(\frac{dx^1}{ds}\right)^2\right] = 0 \end{aligned} \tag{4.82}$$

describe the growth of a two-coral community of sclaractinian corals [Ant90], [AK84], [AFK87], [AL90]. According to the theorem on p. 209 of [AIM93], the Finsler space with Lagrangian (4.81) is a Berwald space with locally constant coefficients. In this case there is a Riemannian metric

$$g_{ij}(x) = \delta_{ij}\exp\left[2(\alpha_1 - \alpha_2 p)x^1 + 2(\alpha_2 + \alpha_1 p)x^2\right],$$

whose geodesic equations coincide with (4.82).

We employ the above system to construct a model of growth and development in the presence of noise and demonstrate that it gives rise to the canalization of biomasses around their 'targeted' values, a phenomenon discovered experimentally by Medawar in his study of the growth of embryonic chicken heart tissue [AB96].

The system will be subject to constant environmental influences (temperature, etc.) expressed mathematically by introducing constants on the right-hand sides of (4.82), which becomes

$$\begin{aligned}\frac{d^2x^1}{ds^2} + 2\tilde{\alpha}_2\frac{dx^1}{ds}\frac{dx^2}{ds} + \tilde{\alpha}_1\left[\left(\frac{dx^1}{ds}\right)^2 - \left(\frac{dx^2}{ds}\right)^2\right] &= -a\tilde{\alpha}_1,\\ \frac{d^2x^2}{ds^2} + 2\tilde{\alpha}_1\frac{dx^1}{ds}\frac{dx^2}{ds} + \tilde{\alpha}_2\left[\left(\frac{dx^2}{ds}\right)^2 - \left(\frac{dx^1}{ds}\right)^2\right] &= -a\tilde{\alpha}_2,\end{aligned} \tag{4.83}$$

where

$$\tilde{\alpha}_1 = \alpha_1 - \alpha_2 p, \quad \tilde{\alpha}_2 = \alpha_2 + \alpha_1 p, \quad a > 0.$$

These constants can be obtained from the potential

$$V(x) = \frac{a}{2}\exp\left(2\tilde{\alpha}_i x^i\right)$$

by computing the gradient, $a\tilde{\alpha}^i = \nabla^i V := g^{ij}\partial_j V$. In addition, we standardize the external environmental influence by setting

$$\tilde{\alpha}_1^2 + \tilde{\alpha}_2^2 = (1+p^2)(\alpha_1^2 + \alpha_2^2) = 1.$$

Our method of developmental noise and canalization rests on Nelson's celebrated conservative diffusion theory [Nel67], [Nel85]. Thus, we replace the smooth trajectory $x(s)$ in (4.83) by a Markov diffusion $\xi(s)$ with generator of the form

$$\mathcal{G} = \frac{\nu^2}{2}\Delta + b,$$

where $\Delta := g^{-1/2}\partial_i\left(g^{1/2}g^{ij}\partial_j\right)$ with $g = \det[g_{ij}]$ is the Laplace-Beltrami operator, ν is the noise intensity, and $b = b^i\partial_i$ is a vector field refereed to as the *forward drift.* We denote by $\rho(s,x)$ the probability density of $\xi(s)$ relative to the volume form $dV g^{1/2}dx^1dx^2$. Let $u := \frac{1}{2}\nabla\ln\rho$ be the so-called *osmotic velocity* and $v := b - u$ the *current velocity.*

Following Guerra and Morato [GM83] and [Nel85], we define

$$A := E\int_0^T \left[\frac{1}{2}v_i v^i - \frac{1}{2}u_i u^i - V\right] ds = E\int_0^T \left[\frac{1}{2}b_i b^i - \nabla_i b^i - V\right] ds.$$

The latter expression enables one to consider A as a functional of the drift field b. In place of solutions $x(s)$ to the deterministic system (4.83) we consider diffusions $\xi(s)$ that are *critical pints* for A under variations of the drift field b such that the initial distribution $\rho(0, x)$ is fixed, which we write briefly as

$$\delta A(b) = 0.$$

This criticality conditions implies the *Nelson-Newton law*

$$\frac{dv^i}{ds} + v^j \nabla_j v^i - u^j \nabla_j u^i - \frac{\nu^2}{2}\Delta u^i = -\nabla^i V. \tag{4.84}$$

In addition, we have the continuity equation

$$\frac{\partial \rho}{\partial s} = -\nabla_i(v^i \rho), \tag{4.85}$$

from which one can see that $v^i\rho$ is the *probability current.*

We shall seek stationary solutions of (4.84), that is, those for which the probability current $v^i\rho$ vanishes everywhere. The latter implies that ρ and u are both independent of s and (4.84) becomes

$$\nabla^i\left(\frac{1}{2}u_j u^j + \frac{\nu^2}{2}\nabla_j u^j - V\right) = 0.$$

Therefore, there is a constant λ such that

$$\frac{1}{2}u_j u^j + \frac{\nu^2}{2}\nabla_j u^j - V + \lambda = 0.$$

This can also be written as

$$\left(-\frac{\nu^2}{2}\Delta + V\right)\psi = \lambda\psi, \tag{4.86}$$

where $\psi = \sqrt{\rho}$. Thus, the problem of finding stationary solutions of (4.84) can be reduced to the eigenvalue problem for the operator $-\frac{\nu^2}{2}\Delta + V$.

On applying the coordinate transformation

$$r = \exp\left(\tilde{\alpha}_1 x^1 + \tilde{\alpha}_2 x^2\right), \quad \theta = \tilde{\alpha}_2 x^1 - \tilde{\alpha}_1 x^2,$$

the eigenvalue problem becomes

$$\left(-\frac{\nu^2}{2}\Delta_{\text{polar}} + \frac{ar^2}{2}\right)\psi = \lambda\psi,$$

where

$$\Delta_{\text{polar}} = \frac{\partial^2}{\partial r^2} + \frac{1}{r}\frac{\partial}{\partial r} + \frac{\partial^2}{\partial \theta^2},$$

which is just the eigenvalue problem for a harmonic oscillator in two dimensions. As is well known, this problem has an increasing sequence of eigenvalues

$$\lambda_n = \sqrt{\nu^2 a}(n+1), \quad n = 0, 1, 2, \ldots .$$

The eigenfunction corresponding to the smallest eigenvalue λ_0 is

$$\psi_0 = C_0 \exp\left(-\frac{1}{2}\sqrt{\frac{a}{\nu^2}}r^2\right),$$

where C_0 is a normalizing constant. For the second eigenvalue λ_1 there are two independent eigenfunctions

$$\psi_{10} = C_{10} x \exp\left(-\frac{1}{2}\sqrt{\frac{a}{\nu^2}}r^2\right),$$
$$\psi_{01} = C_{01} y \exp\left(-\frac{1}{2}\sqrt{\frac{a}{\nu^2}}r^2\right),$$

where $x = r\cos\theta$ and $y = r\sin\theta$ are the Cartesian coordinates. For the third eigenvalue λ_2 there are three linearly independent eigenfunctions, and so on. The corresponding densities $\rho = \psi^2$ provide stationary solutions of (4.84), see Figures 1–4.

In general, a stationary solution ρ has nodal lines $\rho = 0$, which split the plane into disjoint domains. No sample path $\xi(s)$ of the corresponding Nelson diffusion can ever cross any of these lines (i.e. the event has probability zero). The diffusion $\xi(s)$ will therefore remain forever in one of these disjoint regions. Moreover, its is a basic result of Nelson's theory that the sample paths $\xi(s)$ converge to the deterministic solutions of the Euler-Lagrange equations (4.83) with potential V. The confinement of $\xi(s)$ in one of the regions bounded by nodal lines can be interpreted as a mathematical manifestation of the canalization of biomasses in the presence of *developmental noise*, in the sense of C.H. Waddington [AB96].

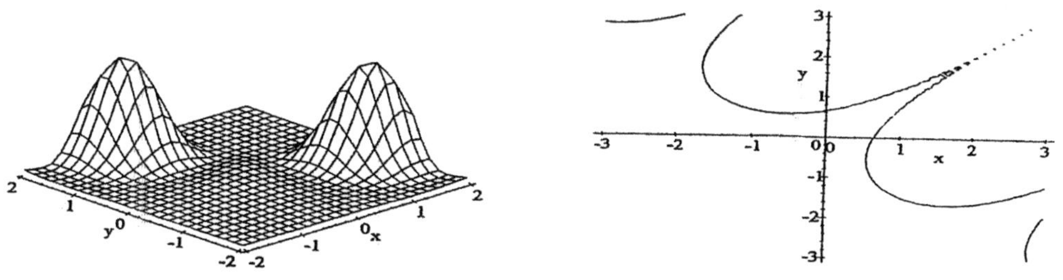

Figure 1: Density $\rho_{02}(x, y)$ and nodal lines $\rho_{02}(x, y) = 0$.

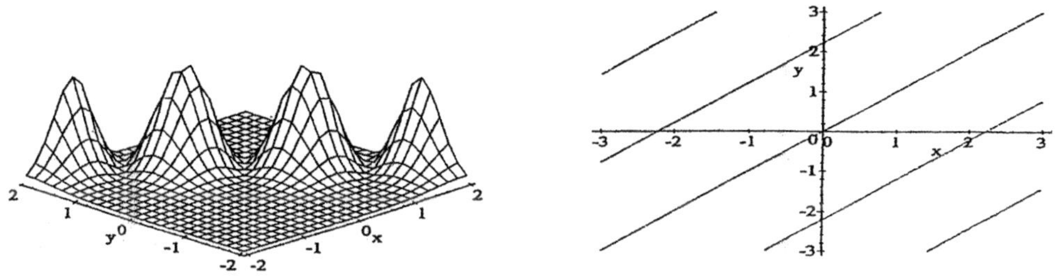

Figure 2. Density $\rho_{11}(x, y)$ and nodal lines $\rho_{11}(x, y) = 0$.

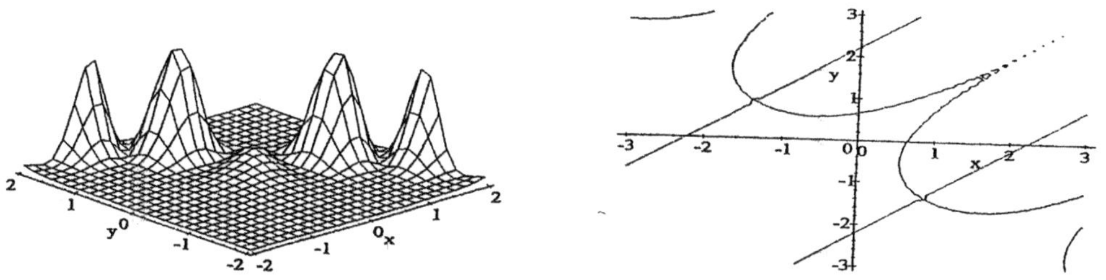

Figure 3: Density $\rho_{12}(x, y)$ and nodal lines $\rho_{12}(x, y) = 0$.

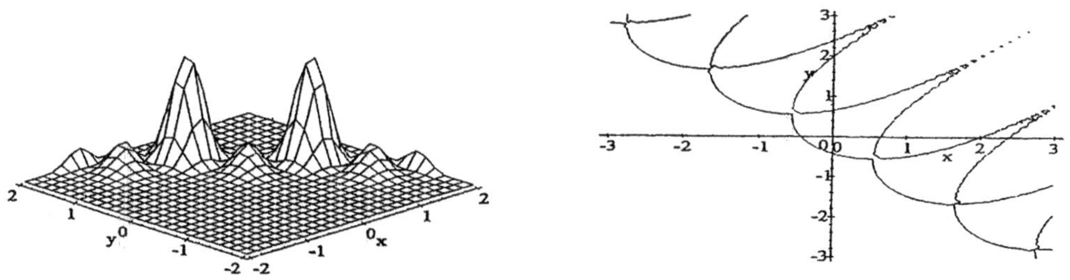

Figure 4: Density $\rho_{22}(x, y)$ and nodal lines $\rho_{22}(x, y) = 0$.

4.8 Noisy Systems in Chemical Ecology and Epidemiology

Our goal is to introduce noise into the Berwald space with metric (4.81) and into the geodesic dynamics of the Finsler metric $\overline{L} = e^{\sigma(x)} \cdot L$, with L being the mth-root Minkowski metric (4.42) We will do this in a way which results in Markov Diffusions on $\mathring{T}M$, the slit tangent bundle over M, with M being diffeomorphic to $\mathbb{R}^2$. Here, we interpret the noisy perturbation as due to *external environmental effect* on the two subpopulations ($n = 2$). This is different in spirit than that of Section 4.7, where Nelson's conservative diffusion was used to model *developmental noise* in the sense of C.H. Waddington. The geodesics of the two metrics, therefore, have an interpretation consistent with Volterra-Hamilton Theory as used in ecology [AB96]. For example, (4.81) has been used to model corals of the Great Barrier Reef on the famous problem of starfish devastation [Ant90], while (4.42) has been used to model myxomatosis, the European wild rabbit disease [AS88a], [AS87a], [AZ94], in which the rabbit flea is the disease vector.

In the absence of noise, the dynamics of either (4.81) or (4.42) is expressed on $\mathring{T}M$ by

$$\begin{aligned} dx^i &= y^i ds, \\ \overline{\delta} y^i &= 0, \end{aligned} \tag{4.87}$$

where $\overline{\delta} y^i = dy^i + \overline{N}^i_j(x,y)dx^j$, $\overline{N}^i_j$ being the coefficients of the appropriate nonlinear connection used in each of the cases. For (4.81) we use the *Cartan nonlinear connection* while for (4.42) we use the *Wagner nonlinear connection*, $\overline{N}^i_j = \partial_j \sigma y^i$.

The way noise enters the deterministic equations must depend on the current state $x(s), y(s)$ of the two populations and on the geometry of the configuration space (production space) in the neighborhood of $x(s)$. In Finslerian spaces, this geometry depends on $y(s)$, as well, but in a way that's dictated by the Finsler metric functions. First observe that any geodesic of either metric can be obtained by hv-rolling the Finsler space along a curve $(\gamma(s), \eta(s))$ in $\mathbb{R}^2 \times \mathbb{R}^2$, where $\gamma(s) = as + b$ is a straight line and $\eta(s) = c$ is a constant. Here, a, b, c are points in $\mathbb{R}^2$ depending on the initial conditions in the equations for hv-rolling. We now perturb $(\gamma(s), \eta(s))$ by the noise $(w(s), v(s))$ to obtain the process $(\gamma(s) + w(s), \eta(s) + v(s))$ in $\mathbb{R}^2 \times \mathbb{R}^2$, along which we hv-roll the manifold using the same initial conditions. Thus, we obtain the process $(x(s), y(s), z(s))$ on OM, that solves the system of SDE's.

$$\begin{aligned} dx^i &= z^i_j \circ dw^j + y^i ds, \\ \bar{\delta} y^i &= z^i_j \circ dv^j + C^i ds, \\ dz^i_j &= -\overline{F}^j_{k\ell}(x,y) z^\ell_j \circ dx^k - C^i_{k\ell}(y) z^\ell_j \circ \bar{\delta} y^k, \end{aligned} \tag{4.88}$$

where $(\overline{N}^i_j, \overline{F}^i_{jk}, C^i_{jk})$ are the coefficients of the *Cartan connection* for the case of (4.81) and those of the *Wagner connection* for the case of (4.42). For the former $C^i \equiv 0$, while for the Wagner case C^i is given in (4.77). Note that $C^i_{jk} = \overline{C}^i_{jk}$, because the *Cartan torsion tensor* is a conformal invariant so they can each be computed directly from the two Minkowski metrics associated with (4.81) and (4.42)

The projected process $(x(s), y(s))$ is a homogeneous Markov diffusion on $\mathring{T}M$ with generator

$$\mathcal{D} = \frac{1}{2}\bar{g}^{ij}(\bar{\delta}_i\bar{\delta}_j - \overline{F}^k_{ij}\bar{\delta}_k) + \frac{1}{2}\bar{g}^{ij}(\dot{\partial}_i\dot{\partial}_j - C^k_{ij}\dot{\partial}_k) + C^i\dot{\partial}_i + y^i\bar{\delta}_i, \tag{4.89}$$

where $\bar{g}_{ij} = e^{2\sigma} g_{ij}$ is the metric tensor of $\overline{L}$ in either (4.81) or (4.42)

Because y^i(s) are interpreted as relative population sizes, the process $(x(s), y(s))$ must be restricted to the region where both y-coordinates are positive, i.e. the process must start at a point (x_0, y_0) with $y^i_0 > 0$ for $i = 1, 2$, and will be considered up to the first moment τ of hitting the set $y^1 y^2 = 0$.

Let $G_{\alpha\beta}(\alpha, \beta = 1, 2, 3, 4)$ be the Riemannian metric tensor on $\mathring{T}M$ obtained by the diagonal lift of g_{ij} $(i, j = 1, 2)$ and let

$$g = \det(g_{ij}), \quad G = \det(G_{\alpha\beta})$$

which means, $g^2 = G$. Denote by $\rho(s, x, y)$ the probability density of the process with generator (4.89) stopped at time τ, to be in a region $A \subseteq \mathring{T}M \cap \{y^1 > 0,\, y^2 > 0\}$ relative to the measure $\sqrt{G}\, dxdy$ in $\mathring{T}M$, i.e.

$$P\{(x(s), y(s)) \in A\} = \int_A \rho(s, x, y)\, \sqrt{G(y)}\, dxdy.$$

The function $\rho(s, x, y)$ satisfies the forward initial boundary value problem

$$\begin{aligned} &\mathcal{D}^* p = \frac{\partial \rho}{\partial s} \\ &\lim_{s \searrow 0} \rho(s, x, y) = \rho_0(x, y) \\ &\rho(s, x, y)\Big|_{y^1 y^2 = 0} = 0, \end{aligned} \tag{4.90}$$

where the initial density $\rho_0(x, y)$ is supported on $\mathring{T}M \cap \{y^1 > 0, y^2 > 0\}$ and $\mathcal{D}^*$ is the adjoint of $\mathcal{D}$ relative to the measure $\sqrt{G}dxdy$.

Let us now proceed to solve (4.90) in each of the two cases, via the Feynman-Kac formula. Let us consider (4.81) first. To this end, we compute $\mathcal{D}^*$ to find

$$\mathcal{D}^* f = \mathcal{D}f - 2y^i\overline{\delta}_i f - 4\overline{g}^{ij}\sigma_i\overline{\delta}_j f + \overline{g}^{ij}C^k_{ik}\dot{\partial}_j f + Vf \qquad (4.91)$$

where use of Theorem 4.13 for the curvature K yields the Feynman-Kac potential,

$$\begin{aligned} V &= 2\overline{K} + 2I\,\overline{m}^i\overline{\ell}^j\sigma_{ij} - 4I\overline{m}^i\overline{\ell}^j\sigma_i\sigma_j - 2I^2\overline{m}^i\overline{m}^j\sigma_i\sigma_j \\ &\quad + 8\overline{g}^{ij}\sigma_i\sigma_j + 4y^i\sigma_i - \frac{1}{2}I^2/\overline{L}^2. \end{aligned} \qquad (4.92)$$

Here,

$$\overline{K} = \frac{1}{2}\overline{g}^{ik}\mathbf{R}^\ell_{jk\ell},$$

when $\mathbf{R}^i_{jk\ell}$ is the h-curvature tensor of the Cartan connection. In the special case we have here the main scalar, $I = 2p/\sqrt{p^2+1}$, is a constant function so the last term in the expression for $\overline{K}$ in (4.48) vanishes.

We need now to introduce the auxiliary process $(\xi(s), \eta(s))$ in $\mathring{T}M$ which is an hv-Brownian motion with additional horizontal and vertical drifts

$$b^i = -2y^i - 4\overline{g}^{ij}\sigma_j, \quad b^{\overline{i}} = \overline{g}^{ij}C^k_{jk}. \qquad (4.93)$$

Thus, $(\xi(s), \eta(s))$ satisfies the system of SDEs

$$\begin{aligned} d\xi^i &= z^i_j \circ w^j + b^i(\xi, \eta)ds \\ \overline{\delta}\eta^i &= z^i_j \circ dv^j + b^{\overline{i}}(\xi, \eta)ds \\ dz^i_j &= -\overline{F}^i_{k\ell}(\xi, \eta)z^\ell_j \circ d\xi^k - C^i_{k\ell}(\eta)z^\ell_j \circ \overline{\delta}\,\eta^k. \end{aligned} \qquad (4.94)$$

We claim that $\overline{L}(\xi(s), \eta(s))$ is a Bessel process with index $n = 2$. The proof is similar to the Bessel process result (Theorem 3.12 in Chapter 3), except that drifts b^i and $b^{\overline{i}}$ must be taken into account. But, $y_k b^{\overline{i}}(x, y) = 0$ so the proof of that result can be adapted to our case here. It follows that $\eta(s)$ cannot have explosions, nor can the orthonormal frame $z(t)$. In order to rule out the possibility that $\xi(s)$ might explode, we *assume that $\sigma(x)$ vanishes for $|x| > B$, where B is a large enough real number.* Biologically, this amounts merely to a statement that the model is not applicable if biomass x exceed some a priori given maximum.

With this support condition (4.94) reads

$$\begin{aligned} d\xi^i &= z^i_j \circ dw^j - 2\eta^i ds \\ d\eta^i &= z^i_j \circ dv^j + Q^i(\eta) ds \\ dz^i_j &= -C^i_{k\ell}(\eta) z^\ell_j \circ d\eta^k \end{aligned}$$

where $Q^i = g^{ij} C^k_{jk}$. Since the right-hand sides of these SDEs are independent of ξ and neither $\eta(s)$ nor $z(s)$ has explosions, $\xi(s)$ cannot have explosions either.

We now denote by $\xi_{x,y}(s)$, $\eta_{x,y}(s)$ the solution of (4.94) with initial conditions $\xi(0) = x$, $\eta(0) = y$, $z(0) = z$. Note, incidentally, that the initial condition for the frame $z(s)$ can be omitted since $(\xi(s), \eta(s))$ is a Markov diffusion process on its own. In addition, we denote by $\tau_{x,y}$ the first hitting time at the surface $y^1 y^2 = 0$, by the process $\eta_{x,y}(s)$. Then the solution to (4.90) can be written as the so-called Feynman-Kac formula

$$\rho(s,x,y) = E\Big\{\chi_{\{\tau_{x,y}>s\}} \rho_0\big(\xi_{x,y}(s), \eta_{x,y}(s)\big) \exp\Big[\int_0^s V\big(\xi_{x,y}(u), \eta_{x,y}(u)\big) du\Big]\Big\}. \tag{4.95}$$

There are a number of conclusions about the qualitative behaviour of $\rho(s,x,y)$ one can draw from this formula. For example, large positive curvature K will cause ρ to increase locally, for the short time asymptotics, while just the opposite will happen in regions which have large negative curvature. In models of toxic soft corals competing for space against reef building hard corals [AS92], [AB96], $\sigma = \alpha_i x^i + \frac{1}{2}\nu(x^1)^2 - \mu x^1 x^2$ and p in (4.81) has the interpretation of a *morphological adaptation* in the hard coral species, to fend off the poisonous soft corals' encroachment. If $\mu > 0$ is not too large, then K is negative so that over the short-run the phase space transition density $\rho(s,x,y)$ is relatively diminished. This parameter is soft-coral *encrouchment,* [ibid].

This is consistent with observations made on Pandora Reef on the Great Barrier Reef of Australia, by Dr. P.W. Sammarco, [AS92], [AB96].

Now let us consider the rabbit/flea model for myxomatosis using for the basic deterministic model the geodesics of metric $\overline{L}$ in (4.42) with a certain linear perturbation term to be described below. We must first, however, develop a notation suitable to this fairly complicated model. An introduction to the deterministic theory can be found in, [AS88a], [AZ94], [AB96].

Coevolution of the rabbit host and flea parasite entails an uncertain environment. Certainly, a portion of the parasite's environment is the host population undergoing random fluctuations. These noisy effects can be added in an intrinsic manner directly from the deterministic dynamics, i.e. from the metric structure of the physiological space, spanned by $x(t) = (x^P(t), x^H(t))$ which denote the (log) total caloric intake up to time t, for the parasite and host populations, prescribed by their ecological interactions. The metric (4.42) is now written

$$\overline{L}^2 = e^{2\phi} \cdot \left[(y^P)^m + (y^H)^m\right]^{2/m}, \quad \phi = \gamma_P x^P - \gamma_H x^H \tag{4.96}$$

and the deterministic equations are

$$\begin{aligned} dx^i &= y^i dt \\ dy^i &= E^i dt - \Gamma^i_{jk}(x, y) y^j y^k \end{aligned} \tag{4.97}$$

with $i, j, k \in \{P, H\}$, $E^P = -\varepsilon_P y^P$, $E^H = -\varepsilon_H y^H$.

Here, $\Gamma^i_{jk}(x, y)$ are the Christoffel symbols for the Finsler metric (4.96) and ε_P and ε_H are metabolic rates (greater for the parasite). The system (4.97) can be written more explicitly as

$$\begin{aligned} \frac{dx^P}{dt} &= y^P, \quad \frac{dx^H}{dt} = y^H \\ \frac{dy^P}{dt} &= -\varepsilon_P y^P - \gamma_P (y^P)^2 + \gamma_H \left(\frac{m}{m-1}\right) y^P y^H + \frac{\gamma_P}{m-1} \left(\frac{y^H}{y^P}\right)^{m-2} (y^H)^2 \\ \frac{dy^H}{dt} &= -\varepsilon_H y^H + \gamma_H (y^H)^2 - \gamma_P \left(\frac{m}{m-1}\right) y^P y^H - \frac{\gamma_H}{m-1} \left(\frac{y^P}{y^H}\right)^{m-2} (y^P)^2. \end{aligned} \tag{4.98}$$

Note that the linear terms will disappear under a transformation to a new time parameter if $\varepsilon_P = \varepsilon_H$, only. Thus, (4.98) are not generally geodesics. In any case, the metric remains biologically relevant, for it captures the ability of the host/parasite co-evolved system to respond to external perturbations from the environment. Namely, a *state, which is represented by a point in the physiological space, M, will be displaced a distance* (in terms of the metric) *proportional to the magnitude of the perturbation.* This is our *noise ansatz.*

The physiological manifold M with metric $\overline{L}$ has the structure of a Wagner space with connection $(\overline{N}^i_j, \overline{F}^i_{jk}, C^i_{jk})$ where $\overline{N}^i_j = \phi_j y^i$, $\overline{F}^i_{jk} = \phi_j \delta^i_k$ and $C^i_{jk} = \frac{1}{2} g^{i\ell} \dot{\partial}_\ell g_{jk}$. We shall see that the Wagner connection, although deflection-free and h- and r-metrical relative to $\overline{L}$, it has non-vanishing torsion $\overline{T}^{i}_{jk} = \phi_j \delta^i_k - \phi_k \delta^i_j$. The equations (4.98) can be written in terms of the Wagner connection as follows

$$\begin{aligned} dx^i &= y^i dt, \\ \overline{\delta} y^i &= \overline{g}^{ij} \overline{T}^{\ell}_{jk} \overline{g}_{\ell m} y^k y^m dt + E^i dt, \end{aligned} \tag{4.99}$$

where,

$$\overline{\delta} y^i = dy^i + \overline{N}^{i}_{j} dx^j, \tag{4.100}$$

and the perturbed equations have the form

$$\begin{aligned} dx^i &= y^i dt + du^i \\ \overline{\delta} y^i &= \overline{g}^{ij} \overline{T}^{\ell}_{jk} \overline{g}_{\ell m} y^k y^m dt + E^i dt + ds^i \end{aligned} \tag{4.101}$$

where,

$$\overline{\delta} y^i = dy^i + \overline{N}^{i}_{j} \circ dx^j. \tag{4.102}$$

The *noise terms* du^i and ds^i *must transform like vectors* and the *Stratonovich circle notation* must be used to ensure that the *stochastic differential equation* (4.101) *are covariant.*

According to our general *noise ansatz*, the perturbation terms du^i and ds^i must be determined by the metric. Because the fibers of $\mathring{T}M$ are non-Euclidean for $m \geq 3$ (each fiber is isometrically isomorphic to a Minkowski space with the mth root metric), the perturbations ds^i added to the so-called population equations are not simple white noises. In order to find the expressions for du^i and ds^i in terms of $\overline{L}(x, y)$, we use hv-rolling of the Finsler space along a curve as introduced in Chapter 3, but this time it will be controlled by the Wagner connection. The result is (4.88) with C^i given by (4.77). Equivalently, we can write

$$\begin{aligned} dx^i &= y^i dt + z^i_j \circ dw^j, \\ \overline{\delta} y^i &= \overline{g}^{ij} \overline{T}^{\ell}_{jk} \overline{g}_{\ell m} y^k y^m dt + E^i dt + z^i_j \circ dv^j, \\ dz^i_j &= -\overline{F}^{i}_{k\ell} z^{\ell}_j \circ dx^k - C^i_{k\ell} z^{\ell}_j \circ \overline{\delta} y^k. \end{aligned} \tag{4.103}$$

Noting that

$$\begin{aligned} du^i &= z^i_j \circ dw^j = z^i_j dw^j + \frac{1}{2} dz^i_j dw^j \\ &= z^i_j dw^j - \frac{1}{2}\overline{F}^i_{k\ell} z^\ell_j dx^k dw^j \\ &= z^i_j dw^j - \frac{1}{2}\overline{F}^i_{k\ell} z^k_m z^\ell_j \delta^{mj} dt \\ &= z^i_j dw^j - \frac{1}{2}\overline{g}^{k\ell}(x,y)\overline{F}^i_{k\ell} dt, \end{aligned}$$

we conclude that

$$u^i(t) - u^i(0) - \frac{1}{2}\int_0^t \overline{g}^{jk}(x(s),y(s))\overline{F}^i_{jk} ds, \tag{4.104}$$

is a *martingale*.

The Markov process (4.103) can now be shown to have generator (4.89) or equivalently,

$$\begin{aligned} \mathcal{D} &= \frac{1}{2}\overline{g}^{jk}(\overline{\delta}_j\overline{\delta}_k - \overline{F}^i_{jk}\overline{\delta}_i) + \frac{1}{2}\overline{g}^{jk}(\dot\partial_j\dot\partial_k - C^i_{jk}\dot\partial_i) \\ &\quad + y^i\overline{\delta}_i + \overline{g}^{ij}\overline{T}^\ell_{jk}\overline{g}_{\ell m} y^k y^m \dot\partial_i + E^i\dot\partial_i. \end{aligned} \tag{4.105}$$

We can compute the adjoint operator to be

$$\mathcal{D}^* = \mathcal{D} + A^i\overline{\delta}_i + B^i\dot\partial_i - 2E^i\dot\partial_i + V \tag{4.106}$$

where

$$\begin{aligned} A^i &= -3\overline{g}^{ij}\phi_j - 2y^i, \\ B^i &= \overline{g}^{ij}C^k_{kj} - 2\overline{L}^2\overline{g}^{ij}\phi_j + 2\phi_j y^j y^i, \\ E^i &= -\varepsilon y^i - \widetilde{y}^i/2\xi, \end{aligned} \tag{4.107}$$

and

$$V = 6\overline{g}^{ij}\phi_i\phi_j - I_\theta/2\overline{L}^2 + I^2/2\overline{L}^2 + 5y^i\phi_i + 2\varepsilon + \widetilde{y}^i C^j_{ji}/\xi, \tag{4.108}$$

where $\widetilde{y}^P = y^P$, $\widetilde{y}^H = -y^H$, $\varepsilon = \frac{1}{2}(\varepsilon_P + \varepsilon_H)$ is the average metabolic rate, and $\xi = 1/\varepsilon_P - \varepsilon_H$ is the so-called efficiency of the system, [AS88a], [AB96]. Here, θ is the *Landsberg angle* (4.44) and I the *main scalar* (4.43).

Once again we introduce an auxiliary process (X, Y) on $\mathring{T}M$ defined by

$$\begin{aligned} dX^i &= Z^i_j \circ dw^j + A^i dt, \\ \overline{\delta} Y &= Z^i_j \circ dv^j + B^i dt - 2E^i dt, \\ dZ^i_j &= -\overline{F}^i_{k\ell} Z^\ell_j \circ dX^k - C^i_{k\ell} Z^\ell_j \circ \overline{\delta} Y^k, \end{aligned} \tag{4.109}$$

where w and v are independent standard Brownian motions on $\mathbb{R}^2$ with respect to a fixed probability measure P.

We denote by $(X_{x,y}, Y_{x,y})$ the Markov diffusion on $\mathring{T}M$ defined by (4.109) with initial conditions $X(0) = x$, $Y(0) = y$ where $(x, y) \in TM^+$ (all coordinates of x, y are positive). Moreover, let $\tau_{x,y}$ be the first hitting time for the boundary $\partial(TM^+)$ by the sample paths of $(X_{x,y}, Y_{x,y})$. Then the solution $\rho(t, x, y)$ to (4.90) in the case at hand can be written as the Feynman-Kac formula

$$\rho(t,x,y) = E\Big\{\chi_{\{\tau_{x,y}>t\}}\rho_0\Big(X_{x,y}(t), Y_{x,y}(t)\Big) \exp\Big[\int_0^t V(X_{x,y}(s), Y_{x,y}(x))ds\Big]\Big\}, \tag{4.110}$$

where E is the expectation relative to P and χ is the indicator function of a set.

We will now use (4.110) to obtain information about the myxomatosis epizootic. We demonstrate how to study the dependence on the average metabolic rate ε and the efficiency ξ of the system. We will find strong relationships between density dependence of the host/parasite system and Finsler geometric quantities.

First, we extract the dependence of the Feynman-Kac potential V in (4.108) on ε and ξ, by use of the Cameron-Martin-Girsanov Theorem. We set

$$\psi^i = 2\zeta^i_j E^j, \quad d\widetilde{w}^i = dw^i, \quad d\widetilde{v}^i = dv^i - \psi^i dt \tag{4.111}$$

where ζ^i_j is the frame dual to Z^i_j. Then the auxiliary diffusion $(X_{x,y}, Y_{x,y})$ defined by (4.109) satisfies the following SDEs:

$$\begin{aligned} dX^i &= Z^i_j \circ d\widetilde{w}^j + A^i dt \\ \overline{\delta} Y^i &= Z^i_j \circ d\widetilde{v}^j + B^i dt \\ dZ^i_j &= -\overline{F}^i_{k\ell} Z^\ell_j \circ dX^k - C^i_{k\ell} Z^\ell_j \circ \overline{\delta} Y^k, \end{aligned} \tag{4.112}$$

where, by the Cameron-Martin-Girsanov Theorem, $\widetilde{w}$, $\widetilde{v}$ are independent standard Brownian motions on $\mathbb{R}^2$ with respect to $\widetilde{P}$ defined by

$$d\widetilde{P} = \exp\Big(\int_0^t \delta_{ij}\psi^i dw^j - \frac{1}{2}\int_0^t \delta_{ij}\psi^i\psi^j ds\Big)dP. \tag{4.113}$$

The Feynman-Kac formula can now be written

$$\begin{aligned}\rho(t,x,y) &= \widetilde{E}\Big\{\chi_{\{\tau_{x,y}>t\}}\rho_0\big(X_{x,y}(t),Y_{x,y}(t)\big)\exp\Big[-\int_0^t \delta_{ij}\psi^i dw^j \\ &\quad +\frac{1}{2}\int_0^t \delta_{ij}\psi^i\psi^j ds+\int_0^t V\big(X_{x,y}(s),Y_{x,y}(s)\big)ds\Big]\Big\}. \qquad (4.114)\end{aligned}$$

The dependence on ε and ξ is now only in the exponent in (4.114) because by (4.112) the probability law of $(X_{x,y},Y_{x,y})$ relative to the new probability measure $\widetilde{P}$ is independent of ε and ξ. From (4.111) we compute

$$\begin{aligned}\delta_{ij}\psi^i dw^j-\frac{1}{2}\delta_{ij}\psi^i\psi^j dt &= 2\,\overline{g}_{k\ell}E^k\circ d\overline{Y}^\ell-2\,\overline{g}_{k\ell}E^kB^\ell dt \\ &\quad +2\,\overline{g}_{k\ell}E^kE^\ell dt-\nabla^v_kE^k dt \qquad (4.115)\end{aligned}$$

where $\nabla^v_jA^i=\dot{\partial}_jA^i+C^i_{jk}A^k$ is the vertical covariant derivative. Thus, substitution for B^i and E^i from (4.107) and using the expression for V, it is possible to write the entire exponent above in the form

$$\begin{aligned}\exp\Big\{2\varepsilon\int_0^t d\overline{L} &\quad +\frac{1}{\xi}\int_0^t \overline{g}_{k\ell}\widetilde{Y}^k\circ\overline{\delta}Y^\ell \\ &\quad +\int_0^t\Big[\frac{2}{\xi}\overline{L}^2\phi_k/\xi-\frac{2}{\xi}\phi_iY^jY^\ell\overline{g}_{k\ell}/\xi-\frac{C^i_{ki}}{2\xi}\Big]\widetilde{Y}^k ds \\ &\quad -\int_0^t 2\overline{g}_{k\ell}\Big(\varepsilon Y^k+\frac{\widetilde{Y}^k}{2\xi}\Big)\Big(\varepsilon Y^\ell+\frac{\widetilde{Y}^\ell}{2\xi}\Big)ds \\ &\quad +\int_0^t\Big[6\overline{g}^{ij}\phi_i\phi_j-\frac{I_\theta}{2\overline{L}^2}+\frac{I^2}{2\overline{L}^2}+5Y^i\phi_i\Big]ds\Big\}.\end{aligned}$$

From this expression we can derive important consequences. Suppose $\varepsilon_P\gg\varepsilon_H$. Then ε is large and ξ is small so the above exponent is dominated by

$$-2\int_0^t \overline{g}_{k\ell}(X,Y)(\varepsilon Y^k+\widetilde{Y}^k/2\xi)(\varepsilon Y^\ell+\widetilde{Y}^\ell/2\xi)ds$$

which is negative when Y^P and Y^H are both positive. This means the transition density is degraded so that on average the process of myxomatosis is speeded up, relatively speaking.

On the other hand, considering fixed values of ε_H and ε_P (or equivalently, of ε and ξ) if y^H is small but $y^P\approx C>0$, where C is some constant, then $I^2/2\overline{L}^2$ and $-I_\theta/2\overline{L}^2$ both tend to $+\infty$ as $y^H\to 0$ and $y^P\to C>0$. All the remaining terms can be shown to be bounded, so that these two Finsler terms involving the main scalar I will dominate the exponent. The result is increased probability density for the transition into the vicinity of the $y^H = 0$ axis, as one would expect in the myxomatosis epizootic when the rabbits have mostly died off. Increased transition density into a neighborhood of $y^H = 0$, also means that the stochastic dynamics is slowed down, there. This slow down makes good biological sense for the end of the epizootic. The major point here is that it is due entirely to density-dependent (i.e. purely Finsler with $m \geq 3$) effects, for when $m = 2$, $I \equiv I_\theta = 0$.

The above conclusions are consistent with the Lyapunov stability results for the deterministic model, when $\varepsilon_P \approx \varepsilon_H$, as follows from (4.49).

4.9 Riemannian Nonlinear Filtering

The role of the *scalar curvature,* $\mathcal{R}$, in Graham's formulation of path-integral theories in quantum mechanics and in nonequilibrium statistical thermodynamics is well known and important [Gra77a], [Gra77b]. Recent work by Takahashi and Watanabe has made it possible for mathematicians to appreciate this, as well [TW81]. However, it was Huzurbazzar and Rao who first brought scalar curvature into statistics via the maximum likelihood surface concept [Huz49], [Rao60]. This section may be viewed as an attempt to bridge a gap between these divergent statistical theories. Here, it is demonstrated that $\mathcal{R}$ enters *nonlinear filtering theory for the estimation of a signal process conditioned on information in the observations process.* Although we do not dwell on this here, it is interesting to compare *Fisher's information matrix* with the *Ricci curvature tensor* $\mathcal{R}_{ij}$, defined in Section 4.10. The interested reader is invited to consult [Ant79], [Huz49], [Rao60] on these matters.

The scalar curvature enters filtering theory via explicit formulas for the C^∞–densities of a large class of (adjoint) Zakai equations. This class is defined in terms of the type of *signal processes* we allow. They are *homogeneous Markov diffusions whose noise terms are conformal*, in the sense of differential geometry explained in Section 4.10. The conformal restriction does not apply to the observations processes which are taken to be homogeneous diffusions, also. In fact, the *conformal restriction is no restriction at all for two dimensional Riemannian signals.* We *do require that signal and observational noise are statistically independent*, however.

Generally, it is proved here that *positive* $\mathcal{R}$ *increases* the *value* of C^∞–*densities for the signal estimates*, while *negative* $\mathcal{R}$ *does just the opposite.* This will be seen to appear explicitly in the measure-valued solutions of the nonlinear Kushner equations, as well as in the C^∞–solutions of the adjoint Zakai equations. Important for our results is the previous work of Kunita [Kun90] on the hypoellipticity problem for stochastic partial differential equations, together with that of Hörmander [Hör69]. Especially relevant to our approach is the backward stochastic calculus for Stratonovich integration of SDEs. We believe, however, that the present work is the first to bring scalar curvature into the theory of stochastic differential equations and nonlinear filtering theory.

The Riemannian scalar curvature has significant application in mathematical biology [Ant85], [AS77], [Ant79], [ACV80], [AV83]. Especially interesting is its interpretation in the growth and chemical ecological dynamics of sessile communities like forests and coral reefs [Ant85], where it can be used to estimate *community vigor.* Large negative $\mathcal{R}$–values indicate rapid growth potential and quick (chemical) responses to predation or herbivory [AE86], [AS88b]. Large positive $\mathcal{R}$–values indicate a relative metabolic passivity [Ant85]. Results of the present study applied to dynamics of sessile organisms imply that signals from vigorous communities are more *difficult to estimate* than for the less vigorous, all other things being equal. This seems intuitively correct. However, more interesting is that *under normalized conditions this difficulty increases more than quadratically with the number of species in the community.*

In the last section we compute results for a much-studied model of starfish predation on an n–species community of corals [AE86], [AK84], [AFK87], [AKR+88].

First a brief review of nonlinear filtering following [Ell82], [Kun83]. In Section 4.10, the necessary geometric definitions are given including the *Levi-Cività connection* and the various *curvature tensors* leading to the Riemann scalar curvature $\mathcal{R}$. This invariant is given explicitly for the class of Riemannian geometries known as *locally conformally flat.* All geometries, known to the authors, which arise in biological applications as above, are of this type. Thus, the coral/starfish example uses this expression for $\mathcal{R}$ to obtain the C^∞–density for the related Zakai equation, in Section 4.11. In Chapter 5 we shall consider Finslerian filtering problems.

We suppose given, once and for all, a probability space $(\Omega, \mathcal{F}, P)$ and a complete, right-continuous, filtration $\{\mathcal{F}_t\}$ of sub σ–fields of $\mathcal{F}$, for $t \in [0, T]$. All processes considered will be $\mathcal{F}_t$–adapted. There are two distinguished classes of processes denoted generally $\{X_t\}$ and $\{Y_t\}$ and called *signal* and *observation* processes, respectively. Write $\mathcal{Y}_t = \sigma\{Y_s : s \leq t\}$ for the σ–field generated by a given observation process Y_t. Note that $\mathcal{Y}_t \leq \mathcal{F}_t$ is usually proper inclusion. Denote by

$$\widehat{X}_t = E(X_t|\mathcal{Y}_t) \tag{4.116}$$

the *least squares estimation* of X_t conditioned on $\mathcal{Y}_t$, and call $\widehat{X}_t$ the *mean filter of the process* X_t rel $\mathcal{Y}_t$.

If Y_t can be given in the form

$$Y_t = \int_0^t h_s ds + L_t, \tag{4.117}$$

where $Y_t = (Y_t^1, \ldots, Y_t^m)$, and where $L_t = (L_t^1, \ldots, L_t^m)$, is an m dimensional Wiener process (i.e., Brownian motion), then the process

$$\nu_t = Y_t - \int_0^t \widehat{h}_s ds \tag{4.118}$$

is a $\mathcal{Y}_t$–Brownian motion called the *innovations process.* This process ν_t will be bounded and $\mathcal{F}_t$–measurable $t \in [0, T]$ and represents *information* in the *observations* $\mathcal{Y}_t$ *concerning* X_t.

In general below we shall be interested in a class of homogeneous strong Markov processes of diffusion type Z_t which are strong solutions of Itô type vector stochastic differential equations like

$$dZ_t = k_s ds + \delta(Z_s) dB_s, \tag{4.119}$$

where $\delta(Z)$ is nonsingular $d \times d$ matrix on $\mathbb{R}^d$, bounded away from zero as

$$\|\delta(Z)\|_e \geq \delta > 0, \qquad Z \in \mathbb{R}^d \tag{4.120}$$

(Euclidean norm), and B_s is a d–dimensional standard Brownian motion. Lipschitz conditions are also invoked. Namely,

$$\|\delta(Z_t) - \delta(Z_t')\|_e \leq K \|Z - Z'\|_t^* \tag{4.121}$$

for all $t \in [0, T]$ and K a constant, and where * denotes the supremum of norms over t. Z_t will have initial conditions independent of $\mathcal{F}_0$. We also require

$$E\Big(\int_0^T \|k_s\|^2 ds\Big) < \infty.$$

(For the rest of Chapter 4 we suspend use of the summation convention on repeated upper and lower indices.) We use the symbol ξ_t for the n–dimensional *signal process* X_t governed by

$$d\xi_t = X_0(t, \xi_t)dt + \sum_{j=1}^{n} X_j \circ d\eta_t^j, \tag{4.122}$$

where η_t^j are n separate Brownian motions and $\circ$ denotes Stratonovich integration. We often will suppose the observation process Y_t is m–dimensional and solves

$$dy_s = h_s ds + \sum_{k=1}^{m} \delta_k(y_s) \circ dB_s^k \tag{4.123}$$

for some *initial condition* independent of $\mathcal{F}_0$ and $\mathcal{Y}_0$, and it will be supposed that sufficient smoothness and boundedness conditions are satisfied so that (4.122) and (4.123) define diffusions equivalent to (4.119), so that we may pass back and forth between Itô and Stratonovich theories, at will.

Let B_s be a standard Brownian motion. The filtration $\mathcal{F}_r^t$ is the least complete σ–field for which all random variables $B_u - B_v : r \leq v \leq u \leq t$ are measurable. If $f(r)$, $r \in [0, t]$ is a continuous stochastic process which is $\mathcal{F}_r^t$–measurable for each r, then the *Itô-Backwards integral* is defined using partitions Δ with f evaluated at right, instead of left, endpoints, i.e.

$$\int_s^t f(r)\widehat{d}B_r = \lim_{|\Delta| \to 0} \sum_{k=0}^{n-1} f(t_{k+1})(B_{t_{k+1}} - B_{t_k}).$$

Of course, the *Backward Stratonovich integral* is defined similarly, but with midpoint averaging, i.e.

$$\int_s^t f(r) \circ \widehat{d}B_r = \lim_{|\Delta| \to 0} \sum_{k=0}^{n-1} \frac{1}{2} \left(f(t_{k+1}) + f(t_k)\right)\left(B_{t_{k+1}} - B_{t_k}\right).$$

We shall use Kunita's method of solving forward *sde* and *spde* problems with backward integrals and Feynman-Kac formulas [Kun83].

Letting $\pi_t(dy, \omega)$ denote the *conditional distribution* of the *signal* given the *observation* data $\mathcal{Y}_t$ we write

$$\pi_t(dy, \omega) = P(\xi_t \in dy | \mathcal{Y}_t)(\omega)$$

for $\omega \in \Omega$ and

$$\pi_t(f) = \int_{\mathbb{R}^n} f(x)\pi_t(dx).$$

As in [Ell82], one shows $\pi_t(f)$ satisfies the *Kushner equation*

$$\begin{aligned}\pi_t(f) &= \pi_0(f) + \int_0^t \pi_s\Big(A(s) + \frac{\partial}{\partial s}\Big) f\, ds \\ &\quad + \int_0^t (\delta^{-1}(y_s)[\pi_s(f \cdot h) - \pi_s(f)\cdot \pi_s(h)] \\ &\quad + \pi_s(D_s f), d\nu_s) \end{aligned} \tag{4.124}$$

where

$$\int_0^t (\phi_s, d\nu_s) \equiv \sum_{i=1}^n \int_0^t \phi_s^i d\nu_s^i$$

and

$$A(s)f(x) = \frac{1}{2}\sum_{j=1}^n X_j^2(s)f(x) + X_0(s)f(x) \tag{4.125}$$

is the infinitesimal generator of the diffusion signal process (4.122). Here we use Hörmander notation so that X_j are $C_b^{1,2}$–vector fields[1] on $\mathbb{R}^n$ with

$$X_j(s)f(x) = \sum_{i=1}^n X_j^i(x)\,\frac{\partial f}{\partial x^i}\,(x).$$

Further,

$$D_s = \big(D_1(s), \ldots, D_m(s)\big),$$

where

$$D_\ell(s)f(x) = \sum_{j=1}^n \gamma_s^{j\ell} X_j(s)f(x); \tag{4.126}$$

$\ell = 1, \ldots, m$ is defined by

$$d\eta_t^j dB_t^\ell = \gamma_t^{j\ell}. \tag{4.127}$$

Here absolute continuity of γ_s is assumed relative Lebesgue measure on $\mathbb{R}^m$ [Ell82]. If we define the α_t–process by

$$\begin{aligned}\alpha_t &= \exp\Big[\int_0^t (\delta^{-1}(y_s)\pi_s(h_s), \delta^{-1}(y_s)dY_s) \\ &\qquad - \int_0^t \|\delta^{-1}(y_s)\pi_s(h_s)\|_e^2 ds\Big].\end{aligned} \tag{4.128}$$

[1]Continuously differentiable in t and twice continuously differentiable in x and the first derivatives are bounded.

Then,

$$\rho_t(f) \equiv \pi_t(f)\alpha_t \tag{4.129}$$

solves the following *Zakai equation* (4.130) provided $\pi_t(f)$ solves the Kushner equation (4.124). The Zakai equation is

$$\begin{aligned} \rho_t(f) &= \rho_0(f) + \int_0^t \rho_s\Big(A(s)f - \frac{1}{2}\sum_{k=1}^m M_k^2(s)f\Big)ds \\ &\quad + \sum_{k=1}^m \int_0^t \rho_s(M_k(s)f) \circ dY_s^k, \end{aligned} \tag{4.130}$$

where $M_k(s) = D_k(s) + h_k(s)$ and $f \in C_b^2(\mathbb{R}^n)$. Also note that

$$\pi_t(f) = \frac{\rho_t(f)}{\rho_t(1)} \tag{4.131}$$

so that $\pi_t(f)$ may be recaptured from the measure solutions of the *linear* Zakai equation. Note that the operator

$$L(s) = A(s) - \frac{1}{2}\sum_{k=1}^m M_k^2(s) \tag{4.132}$$

is nontrivially elliptic [Kun83]. If $\rho_t(dy)$ is a measure solution of (4.130) having a C^∞-density it must satisfy the *adjoint Zakai equation*

$$d\rho_t = L^*(t)\rho_t dt + \sum_{k=1}^m M_k^* \rho_t \circ dY_t^k, \tag{4.133}$$

where

$$L^* = \frac{1}{2}\sum_{i=1}^n X_i^2 + X_0^* + h_0^* \tag{4.134}$$

$$X_0^* = -X_0 + \sum_{j=1}^n \Big(\sum_{i=1}^n \partial_{x^i} X_j^i\Big) X_j, \tag{4.135}$$

$$\begin{aligned} h_0^* &= h_0 + \frac{1}{2}\sum_{j=1}^n \Big(\sum_{i=1}^n \partial_{x^i} X_j^i\Big)^2 - \sum_{i=1}^n \partial_{x^i} X_0^i \\ &\quad + \frac{1}{2}\sum_{j=1}^n \sum_{k=1}^m X_j^k \partial_{x^k}\Big(\sum_{i=1}^n \partial_{x^i} X_j^i\Big) \end{aligned} \tag{4.136}$$

$$M_k^* = -D_k + h_k^*, \qquad k = 1, \ldots, m, \tag{4.137}$$

and, finally,

$$h_k^* = h_k - \sum_{i=1}^{n} \partial_{x^i} D_k^i, \qquad k = 1, \ldots, m. \tag{4.138}$$

The coefficients of the vector fields X_j and D_k are assumed bounded and continuous, C^1 in t, C^4 in $\boldsymbol{x}$, and C_b^1 in $\boldsymbol{x}$.

We are interested in computing C^∞–densities for the *(forward) stochastic pde* (4.130) which is in adjoint form. In order to do this we make use of the *backward Stratonovich calculus* [Kun83]. We will need to solve the *backward equation*

$$\begin{aligned} \widehat{d}\,\widehat{\xi}_s &= -X_0^*(s, \widehat{\xi}_s)ds - \sum_{j=1}^{n} X_j(s, \widehat{\xi}_s) \circ \widehat{d\eta}_s^j \\ &\quad - \sum_{k=1}^{m} D_k(s, \widehat{\xi}_s) \circ \widehat{d}B_s^k, \end{aligned} \tag{4.139}$$

with *terminal condition* $\widehat{\xi}_{t,t}(\boldsymbol{x}) = \boldsymbol{x}$. The unique solution passing through $(\boldsymbol{x}, t)$ is then written in terms of Stratonovich backward integrals as

$$\begin{aligned} \widehat{\xi}_{s,t}(\boldsymbol{x}) &= \boldsymbol{x} + \int_s^t X_0^*\big(r, \widehat{\xi}_{r,t}(\boldsymbol{x})\big)dr \\ &\quad + \sum_{j=1}^{n} \int_s^t X_j\big(r, \widehat{\xi}_{r,t}(\boldsymbol{x})\big) \circ \widehat{d\eta}_r^j \\ &\quad + \sum_{k=1}^{m} \int_s^t D_k(s, \widehat{\xi}_{r,t}) \circ \widehat{d}B_r^k. \end{aligned} \tag{4.140}$$

If we let z be a *real variable* and write

$$\begin{aligned} \widehat{\Phi}_{s,t}(\boldsymbol{x}, z, \omega) &\equiv z \cdot \exp\Big[\sum_{k=1}^{m} \int_s^t h_k^*\big(r, \widehat{\xi}_{r,t}(\boldsymbol{x})\big) \circ \widehat{d}B_r^k \\ &\quad + \int_s^t h_0^*\big(r, \widehat{\xi}_{r,t}(\boldsymbol{x})\big)dr\Big], \end{aligned} \tag{4.141}$$

where h_0^* and h_k^* are given in (4.136) and (4.138), then the C^∞–density for the measure solution of the Zakai equation (4.130) must be

$$\rho_t(\boldsymbol{x}, \omega) = E\big[\rho_0\big(\widehat{\xi}_{t_0,t}(\boldsymbol{x}, \omega)\big) \cdot \widehat{\Phi}_{t_0,t}(\boldsymbol{x}, 1, \omega)\big]. \tag{4.142}$$

We will compute $\rho_t(\boldsymbol{x}, \omega)$ for certain systems in the next section because it will be shown that C^∞–densities exist for these special systems.

4.10 Conformal Signals and Geometry of Filters

We consider signals ξ_t of the form (4.122) for which X_0 and $X_j = e^{-\phi}\partial_j$, $j = 1, \ldots, n$, are C^ω-vector fields (i.e. real analytic) with bounded first and second partials. Also, we suppose

$$d\eta_t^j dB_t^i = 0$$

for the noise functionals of the signal and observations process Y_t. We require Y_t to be a homogeneous strong Markov process solving

$$dy_s = h_s(x)ds + \delta_k(y_s) \circ dB_s^k. \tag{4.143}$$

Also, $h_s(x) = \left(h_s^1(x), \ldots, h_s^m(x)\right)$ are supposed C^ω for all $x \in \mathbb{R}^n$. The operator for the *conformal signal process* ξ_t is

$$A(s) = \frac{1}{2}\sum_{i=1}^{n} e^{-\phi}(\partial_i(e^{-\phi}\partial_i)) + \sum_{i=1}^{n} X_0^i\partial_i - \frac{\partial}{\partial t}\cdot$$

Such a linear differential operator is said to be *hypoelliptic* in an open set $U \subseteq \mathbb{R}^{n+1}$ if every distributional function u in U is C^∞ in every open subset $W \subseteq U$ in which $A(s)u$ is C^∞ [Hör69], p. 148. Thus, if $A(s)$ is hypoelliptic with distributional solution u of $A(s)u = 0$, it follows that u can be modified on a set of measure zero to be C^∞.

Noting that $M_k = h_k$ because of (4.143), it follows that the operator $L(s)$ of (4.132) is C^ω–elliptic, with scalar term

$$h_0 = -\frac{1}{2}\sum_{k=1}^{m} h_k^2,$$

if $A(s)$ is C^ω–elliptic. Also, the adjoint operator, $L^*(s)$, has $h_k^* = h_k$ from (4.138) and

$$h_0^* = h_0 - \sum_i \partial_i X_0^i + \frac{1}{2}e^{-2\phi}\sum_{j=1}^{n}[2(\partial_j\phi)^2 - \partial_j^2\phi] \tag{4.144}$$

follows directly from (4.136) and the form of $A(s)$ above. It is clear that $L^*(s)$ is C^ω–elliptic, so by a theorem of Hörmander the dimension of the Lie algebra generated by $\{X_0 - \partial/\partial t, X_1, \ldots, X_n\}$ is maximal and equal to $n+1$ [Hör69], ibid., [Kun83], p. 150. Since C^ω, implies C^∞, the result of Kunita ensures the existence of a C^∞–density for the measure solution $\rho_t(f)$ of the

Zakai equation (4.130). Recall that $D_k \equiv 0$, here. This density is given by (4.141) and (4.142). Direct substitution of h_k for h_k^* and (4.144) gives the density explicitly.

We wish to describe the Zakai C^∞–density geometrically. Therefore, we digress into some background material on *locally conformally flat Riemannian spaces*, which is basic to the type of signal processes we use in our main results.

Let us rewrite the signal process (4.122) as

$$(d\xi_t^i \equiv) dx_t^i = \sum_{k=1}^{n} \sigma_k^i(\boldsymbol{x}_t) \circ d\eta_t^k + X_0^i dt.$$

The corresponding infinitesimal generator for the Markov process solution ξ_t is the elliptic operator

$$\begin{aligned} A(s) &= \frac{1}{2}\sum_{i,j}^{n} g^{ij}(\boldsymbol{x}_s)\partial_i\partial_j - \frac{\partial}{\partial t} + \sum_{i}^{n} X_0^i \partial_i \\ &\quad - \frac{1}{2}\sum_{r,k}^{n} (\partial_k \sigma_r^m(\boldsymbol{x}_s)) \cdot \sigma_r^k(\boldsymbol{x}_s)\partial_m, \end{aligned} \tag{4.145}$$

where

$$g^{ij} = \sum_{k=1}^{n} \sigma_k^i \cdot \sigma_k^i \tag{4.146}$$

and

$$ch^m = \frac{n-2}{2}\sum_{r,k}(\partial_k \sigma_r^m) \cdot \sigma_r^k \tag{4.147}$$

is the so-called *Stratonovich term* or *Christoffel field* [AS77], [Ant79], [ACV80].

Both g_{ij} and its inverse g^{ij} are second-order *tensor fields* on a *Riemannian space* $(\mathbb{R}^n, g_{ij})$ whose arc-length element is given by

$$ds^2 = \sum_{i,j}^{n} g_{ij} dx^i \otimes dx^j$$

(see [Eis49] or [Kry68]). A metric in the usual sense is obtained (1.50) by minimizing

$$\int_{t_0}^{t_1} \sqrt{\left(\sum_{i,j} g_{ij}\frac{dx^i}{dt}\frac{dx^j}{dt}\right)}\, dt$$

over all C^{∞}−curves γ joining $\boldsymbol{x}_0 = \gamma(t_0)$ to $\boldsymbol{x}_1 = \gamma(t_1)$. This definition can be shown independent of the particular parameter used. If $\boldsymbol{x}_0$ and $\boldsymbol{x}_1$ are close enough in this metric then the minimizing curve γ is a solution to the *geodesic equations*

$$\frac{d^2x}{ds^2} + \sum_{j,k}^{n} \Gamma^i_{jk} \frac{dx^j}{ds} \cdot \frac{dx^k}{ds} = 0,$$

$i, j, k = 1, \ldots, n$. Here, the Riemannian length of dx/ds, is unity, that is, $d\gamma/ds$ is a unit vector, and the solution starts at $\boldsymbol{x}_0$ and ends at $\boldsymbol{x}_1$. The 3-index symbol Γ^i_{jk} plays a fundamental role here, in that it defines *Riemannian parallelism* on the one hand, and the *Stratonovich term*, or *Christoffel field*, on the other. Given the Riemannian metric tensor g_{ij}, the *Levi-Cività connection* is (1.55)

$$\Gamma^i_{jk} = \frac{1}{2} \sum_{\ell=1}^{n} g^{i\ell}(\partial_k g_{j\ell} + \partial_j g_{k\ell} - \partial_\ell g_{jk}). \tag{4.148}$$

Note that Γ^i_{jk} all vanish if $g_{ij} = \delta_{ij}$, the identity matrix, but that a smooth change of coordinates need not preserve this nullity because the Γ do *not* constitute a tensor. Nevertheless, the geodesic equations are invariant under the transformation law of the connection Γ [Eis49], [Kry68]. The quantities g_{ij} and Γ^i_{jk} and their first partial derivatives define the basic curvature notions of Riemannian geometry, $\mathcal{R}^i_{jk\ell}, \mathcal{R}_{jk}$, and $\mathcal{R}$. The most important of these is the (full) *Riemannian curvature tensor* (see Section 1.6)

$$\mathcal{R}^i_{jk\ell} = \partial_\ell \Gamma^i_{jk} - \partial_k \Gamma^i_{j\ell} + \sum_{r=1}^{n} \Gamma^r_{jk}\Gamma^i_{r\ell} - \sum_{r=1}^{n} \Gamma^r_{j\ell}\Gamma^i_{rk}. \tag{4.149}$$

It is basic that $\mathcal{R}^i_{jk\ell} \equiv 0$ for all i, j, k, ℓ, if and only if g_{ij} is transformable to the Euclidean metric tensor δ_{ij}. But, this is only a *local statement.* It holds for $S^1 \times \mathbb{R}^1$, the ordinary straight cylinder, for example.

The next most important curvature is the so-called *Ricci curvature*

$$\mathcal{R}_{jk} \equiv \sum_{s=1}^{n} \mathcal{R}^s_{jks}. \tag{4.150}$$

This tensor is powerful enough to distinguish (locally) between all 2- or 3-dimensional Riemannian spaces, as is $\mathcal{R}^i_{jk\ell}$ [Eis49], [Kry68]. Finally, the *Riemann scalar curvature*,

$$\mathcal{R} \equiv \sum_{j,k} g^{jk}_{jk}, \tag{4.151}$$

is good enough to distinguish all (2-dimensional) Riemann surfaces, locally. All such surfaces are *locally conformally flat*, in that it is possible to use the Cauchy-Riemann equations to prove the existence of so-called *isothermal coordinates* u^1 and u^2 so that

$$g_{ij} = e^{+2\phi}\delta_{ij}, \qquad \phi = \phi(u^1, u^2)$$

[Kry68]. The Riemann scalar curvature is then

$$\mathcal{R} = -2e^{-2\phi}(\partial_1^2\phi + \partial_2^2\phi).$$

Note that there is *always* lurking behind $\mathcal{R}^i_{jk\ell}$, $\mathcal{R}_{jk}$, and $\mathcal{R}$, a *sign convention.* It is simply a matter of how one arranges the minus signs in (4.149). In [Eis49], the sign convention is opposite to that used here, for example. This carries over systematically to $\mathcal{R}_{jk}$ and $\mathcal{R}$.

In general, n-dimensional Riemannian spaces for which

$$g_{ij} = e^{2\phi}\delta_{ij} \tag{4.152}$$

holds in some coordinate patch are called *locally conformally flat spaces.* Their Levi-Cività connections are then given (see [Ant85]) as

$$\begin{aligned} \Gamma^i_{ii} &= \partial_i\phi \\ \Gamma^i_{ji} &= \Gamma^i_{ij} = \partial_j\phi, && i \neq j \\ \Gamma^i_{jj} &= -\partial_i\phi, && i \neq j \\ \Gamma^i_{jk} &= 0, && i \neq j \neq k. \end{aligned} \tag{4.153}$$

The Riemann scalar curvature is given by

$$\mathcal{R} = -(n-1)e^{-2\phi}\sum_j [2(\partial_j^2\phi) + (n-2)(\partial_j\phi)^2] \tag{4.154}$$

(see [Eis49]).

Returning to (4.144) and the *conformal signal* X_0, $X_j = e^{-\phi}\partial_j$, use of (4.154) allows us to write

$$h_0^* = h_0 - \sum_i \partial_i X_0^i + \frac{1}{4}\left\{(n+2)\|\text{grad}_g\phi\|_g^2 + \frac{\mathcal{R}}{n-1}\right\}, \tag{4.155}$$

where

$$\|\text{grad}_g\phi\|_g^2 \equiv \sum_i e^{-2\phi}(\partial_i\phi)^2 \tag{4.156}$$

is the *Riemannian norm* of the *Riemannian gradient* of ϕ,

$$\mathrm{grad}_g^i \phi \equiv \sum_j g^{ij} \partial_j \phi = e^{-2\phi}(\partial_i \phi). \tag{4.157}$$

It now follows that the $\widehat{\Phi}$–function of (4.141) is

$$\begin{aligned} \widehat{\Phi}_{s,t}(\boldsymbol{x}, z, \omega) &= z \cdot \exp\Big\{ \int_s^t \Big(h_0 - \sum_i \partial_i X_0^i \\ &\quad + \frac{1}{4}\Big[(n+2)\|\mathrm{grad}_g \phi\|_g^2 + \frac{\mathcal{R}}{n-1}\Big]\Big)_{(r,t)} dr \\ &\quad + \sum_{k=1}^m \int_s^t h_k\big(r, \widehat{\xi}_{r,t}(\boldsymbol{x})\big) \circ \widehat{d}B_r^k \Big\}, \end{aligned} \tag{4.158}$$

where $\widehat{\xi}_{s,t}$ solves (4.139) with $\widehat{\xi}_{t,t} = \boldsymbol{x}_t$ and $D_k \equiv 0$. It is clear that for *conformal signals, regarding $\mathcal{R}$ as an independent variable, if $\mathcal{R}$ is generally positive, $\widehat{\Phi}_{s,t}$ has a relatively larger value, while if $\mathbb{R}$ is negative generally, $\widehat{\Phi}$ has a relatively smaller value.* This carries over to the C^∞–density of the Zakai equation. Furthermore, because

$$\rho_t(f) = \int_{\mathbb{R}^n} f(\boldsymbol{x})\rho_t(\boldsymbol{x})d\boldsymbol{x}$$

and

$$\pi_t(f) = \frac{\rho_t(f)}{\rho_t(1)}$$

solve the Kushner equation, *this behaviour carries over to the estimation problem* for any suitably smooth f, *providing $\alpha_t = \rho_t(1)$ is independent of* $\mathbb{R}$.

The *estimation of $f(\xi_t)$ conditional on observational data from Y_t, is larger if $\mathcal{R}$ is generally positive, and is smaller if $\mathcal{R}$ is generally negative, providing $\rho_t(1)$ is independent of $\mathcal{R}$.*

Remark 4.6. Because any Riemannian metric on a 2-dimensional surface is *locally conformally flat, all two-dimensional signals* of the form (4.122) can be considered to be conformal signals. This firmly establishes the role of $\mathcal{R}$ in the *associated Zakai* and *Kushner* equations for the *nonlinear filtering problems of signal dimension two.*

4.11 Riemannian Filtering of Starfish Predation

We follow [AV83] in this section (but consult [AE86], [AS87a], [AS88a] also). Let $\mathbb{R}^n_{++}$ denote the subspace of $\mathbb{R}^n$ defined by all $x^i > 0$. Define the metric

$$g_{ij} = e^{2(\sum_i \alpha_i x^i)}\delta_{ij} \equiv e^{2\phi}\delta_{ij},$$

where all $\alpha_i > 0$ are constants. Extend this metric to all of $\mathbb{R}^n$ by modification with C^∞*–bump functions* as follows. Let $\mathbb{C}$ denote a unit size *collar* of the boundary of $\mathbb{R}^n_{++}$. Define

$$g_{ij} = \begin{cases} e^{2(\sum_i \alpha_i b(x^i) x^i)}\delta_{ij} & \text{on } \overline{R^n_{++}} \\ \delta_{ij} & \text{on } \mathbb{R}^n - \mathbb{R}^n_{++}, \end{cases}$$

where

$$b(x^i) = \begin{cases} \frac{e^{-(1/x^i)}}{e^{-(1/x^i)} + e^{-1/(1-x^i)}}, & 0 < x^i < 1 \\ 0 & \text{for } \; x^i \le 0 \\ 1 & \text{for } \; x^i \ge 1, \end{cases}$$

and all i. Note $\partial_i b(x^i) > 0$ for $0 < x^i < 1$ and all orders of derivatives are bounded. Thus, $\widehat{g}_{ij} = e^{2\widehat{\phi}}\delta_{ij} = e^{2\phi}\delta_{ij} = g_{ij}$ on $\mathbb{R}^n_{++} - \mathbb{C}$ and is the flat metric outside $\mathbb{R}^n_{++}$. Clearly, then $e^{-\widehat{\phi}}$, $\partial_i(e^{-\widehat{\phi}})$, $\partial_{j_1}\ldots\partial_{j_k}e^{-\widehat{\phi}}$ are all bounded in $\mathbb{R}^n$.

It is necessary to discuss the *hypoellipticity* of

$$\widehat{A}(s) = \frac{1}{2}\sum_{i,j}^{n}\widehat{g}^{ij}\Big(\partial_i\partial_j - \sum_k^n \widehat{\Gamma}^k_{ij}\partial_k\Big) - \frac{\partial}{\partial t} + \sum_{i=1}^{n} X_0^i\partial_i \tag{4.159}$$

on $\mathbb{R}^n$ in order to ensure the existence of a C^∞–density for the Zakai equation for the conformal signal (4.161) and observations process (4.163), below. The operator $A(s)$ is clearly elliptic on $\mathbb{R}^n_{++}$ and, by a well-known theorem, supports a C^∞–fundamental solution (i.e., is hypoelliptic). $\widehat{A}(s)$ agrees with $A(s)$ on $\mathbb{R}^n_{++} - \mathbb{C}$, defined in terms of g_{ij} in the same way as $\widehat{A}(s)$ is defined in terms of $\widehat{g}_{ij}$. But, $A(s)$ is C^ω–elliptic. By Hörmander's theorem, this analyticity implies the Lie algebra generated by $\{X_0 - \partial/\partial t, X_1, \ldots, X_n\}$ has maximal rank equal to $n+1$, on $\mathbb{R}^n_{++} - \mathbb{C}$ [Hör69], p. 149, [Kun83], p. 150. Outside $\mathbb{R}^n_{++}$, $\widehat{A}(s)$ has Euclidean symbol and again its associated Lie algebra has rank $n+1$. We claim this is true in the collar $\mathbb{C}$, as well. The argument applied to our special case is due to Hörmander [Hör69], p. 149. He shows that if $\widehat{A}(s)$ is not of maximal rank at $x \in \mathbb{C}$, then the Frobenius theorem can be used to show that $\widehat{A}(s)$ is not hypoelliptic. But we know that $\widehat{A}(s)u = 0$ has a C^∞–solution which is in fact a density on $\mathbb{R}^n$. Thus, $\widehat{A}(s)$ has maximal rank everywhere.

Now, using a theorem of Kunita [Kun83], p. 153, the C^∞–vector fields $X_0, X_1, \ldots, X_n$, with $D_k \equiv 0$, $k = 1, \ldots, m$, and the maximal rank Lie algebra condition implies the associated Zakai equation has a C^∞–density for its measures solution. The L(s) operator (4.132) for this problem is $\widehat{A}(s) - \frac{1}{2}\sum_{k=1}^m M_k^2(s)$ and will be further described.

Because we are interested in $\widehat{g}_{ij}$ only for biological reasons we may use g_{ij}, ϕ instead of $\widehat{g}_{ij}$, $\widehat{\phi}$ in what follows. Consequently, it is the behaviour in $\mathbb{R}^n_{++}$ which mainly concerns us in this example. The Levi-Cività connection Γ_ϕ is given by (4.153) to be

$$\begin{aligned} &\Gamma^i_{ii} = \alpha_i, && \\ &\Gamma^i_{ji} = \Gamma^i_{ij} = \alpha_j, && i \neq j \\ &\Gamma^i_{jj} = -\alpha_i, && i \neq j \\ &\Gamma^i_{jk} = 0, && i \neq j \neq k \end{aligned} \tag{4.160}$$

(see [Ant85]).

The signal process to be considered is defined initially only on $\mathbb{R}^n_{++} \times \mathbb{R}^n_{++}$ by

$$\xi_t : \begin{cases} dx^i_t &= N^i_t dt + e^{-\phi} d\eta^i_t \\ dN^i_t &= \Big[-\sum_{j,k}^n \Gamma^i_{jk} N^j N^k + \lambda_i N^i - \delta_i N^i F \Big] dt + d\eta^i_t, \end{cases} \tag{4.161}$$

where $\lambda_i > 0$ and $\delta_i > 0$ are $2n$ fixed constants and F is defined on $\mathbb{R}^i_+$.[2] Extension to all of $\mathbb{R}^n \times \mathbb{R}^n$ is done by regarding the tangent bundle $T\mathbb{R}^n$ as identical to $\mathbb{R}^n \times \mathbb{R}^n$, with the base space spanned by $x^1, \ldots, x^n$ and the canonical fiber spanned by $N^1, \ldots, N^n$. We define a Riemannian metric tensor on $T\mathbb{R}^n$ as $ds^2 = e^{+2\phi}\sum_{i=1}^n (dx^i)^2 + \sum_{i=1}^n (dN^i)^2$. In this way $T\mathbb{R}^n$ becomes the *product Riemannian manifold* $(\mathbb{R}^n, \widehat{g}_{ij}) \times (\mathbb{R}^n, \delta_{ij})$. Replacing ϕ by $\widehat{\phi}$ in (4.161) yields the *extended* ξ_t–process, $\widehat{\xi}_t$. Note that $\widehat{\xi}_t$ is *not a conformal signal*, strictly speaking, rather it is an *extended conformal signal.*

[2] δ_i is to be modified by multiplication with a C^∞ bump function so that it is zero for large enough F–values, likewise for γ in (4.163).

The scalar curvature $\mathcal{R}$ of the product metric is numerically identical with that for the g_{ij} factor alone because the canonical fiber $\mathbb{R}^n$ of $T\mathbb{R}^n$ is flat Euclidean. Therefore, the curvature for $\xi_t, \mathcal{R}_\xi$, is

$$\mathcal{R}_\xi = -(n-1)(n-2)e^{-2\phi}\left[\sum_{i=1}^{n}(\alpha_i)^2\right]. \tag{4.162}$$

Note that $|\mathcal{R}_\xi| \downarrow 0$, as $\|x\|_e \uparrow \infty$ for $x \in \mathbb{R}^n_{++}$, and that $\mathcal{R}_\xi$ becomes more negative with increasing n. In fact, it does so at least *quadratically* keeping ϕ constant as n varies. Finally, note that for $n = 2$, $\mathcal{R}_\xi = 0$.

Now the observations process to be considered here is

$$\left.\begin{aligned} (dY_t \equiv)\ dF_t &= \psi(N_t, F_t)F_t dt + dW_t \\ \text{where} \qquad h_t = \psi &= -\varepsilon + \gamma F_t + \textstyle\sum_{i=1}^{n} \delta_i' N_t^i \end{aligned}\right\} \tag{4.163}$$

and $\varepsilon, \gamma, \delta_i'$ are positive constants.[3] We are supposing $d\eta^i dW = 0$ for all i. Here, $m = 1$, $D_1 \equiv 0$, so $M_1 = h_1 = \psi$. Therefore,

$$\sum_{k=1}^{m} \frac{1}{2} M_k^2(s) = \frac{1}{2}\psi_s^2 \quad \text{and} \quad L(s) = \widehat{A}(s) - \frac{1}{2}\psi_s^2.$$

The adjoint Zakai equation is

$$d\rho_t = \left(\widehat{A}(t)^* - \frac{1}{2}\psi_t^2\right)\rho_t dt + (\psi_t \cdot \rho_t) \circ dW_t$$

as follows from (4.132), (4.133), (4.137), (4.138). Note that integration may be taken to be of Itô type. The C^∞–density $\rho_t(x, \omega)$ may now be constructed as in (4.142) using (4.133). Thus, for the *extended conformal signal* (4.161) we have X_0 given by

$$\begin{aligned} X_0^i &= N^i, \qquad i = 1, \ldots, n \\ X_0^{i+n} &= -\sum_{j,k}^{n} \Gamma_{jk}^i N^j N^k + \lambda_i N^i - \delta_i N^i F, \end{aligned}$$

and the *divergence* of X_0 is given by $\sum_{i=1}^{2n} \partial_i X_0^i = \sum_{i=1}^{n}(\partial X_0^{i+n}/\partial N^i)$, because we *suppose x^i and N^i are independent variables.* We leave it to the

[3] δ_i' is modified by multiplication with C^∞ bump function so it is zero for large enough F-values; likewise for γ in (4.163).

interested reader to compute this divergence in more detail from (4.153) and the definition of $\widehat{\phi}$. The $\widehat{\Phi}$−functional can be explicitly written as

$$\begin{aligned}\widehat{\Phi}_{x,t}(x,z,\omega) &= z\cdot\exp\Big\{\int_s^t\Big[-\sum_{i=1}^n\partial_{N^i}X_0^{i+n}\\&+\frac{1}{4}\Big((n+2)\sum_{j=1}^n e^{-2\widehat{\phi}}(\partial_i\widehat{\phi})^2+\frac{\mathcal{R}_\xi}{n-1}\Big)\Big](r,\widehat{\xi}_{r,t})dr\\&-\frac{1}{2}\int_s^t\psi^2(r,\widehat{\xi}_{r,t})dr+\int_s^t\psi(r,\widehat{\xi}_{r,t}(x))\circ\widehat{dW}_r\Big\}.\end{aligned}\tag{4.164}$$

In the region of biological interest, $\mathbb{R}^n_{++}\times\mathbb{R}^N_{++}$, the positive quantity $\sum_i^n e^{-2\widehat{\phi}}(\partial_i\widehat{\phi})^2$ becomes $e^{-2\phi}\cdot\sum_{i=1}^n(\alpha_i)^2$, and $\mathbb{R}_\xi$ is given by (4.162). Also, $\widehat{\xi}_s$ solves the *backward Stratonovich SDE* in $\mathbb{R}^n\times\mathbb{R}^n$,

$$\widehat{d\widehat{\xi}_s}=-X_0^*(s,\widehat{\xi}_s)ds-\sum_{j=1}^n X_j(s,\widehat{\xi}_s)\circ\widehat{d\eta}_s^j,$$

as follows from (4.139). X_0^* is computed from (4.135) replacing n with $2n$ and defining X_j^i in the obvious way from the product metric "square root".

Note that, since the Γ^i_{jk}'s in (4.161) are constants [see (4.160)], so that the second equation (4.161) is independent of x. It follows, in particular, that N and F are independent of the curvature $\mathcal{R}_\xi$, which by (4.162) is a function of x only. Hence, the h_t−process in (4.163) is also independent of $\mathcal{R}_\xi$. It now follows that $\alpha_t=\rho_t(1)$ [see (4.128), (4.129)] is independent of $\mathcal{R}_\xi$, and so, as indicated in the discussion following (4.158), we can conclude that the Kushner measure

$$\pi_t(f)=\rho_t(f)/\rho_t(1)\tag{4.165}$$

increases with increasing distance from the origin in $\mathbb{R}^n_{++}\times\mathbb{R}^n_{++}$, *because* $\mathcal{R}_\xi$ *becomes less negative with distance. Also,* $\mathcal{R}_\xi$ *decreases at least quadratically with n, other things being equal* (see the discussion following (4.162)). Interpretation of this is straightforward. Namely, the "difficulty" in estimating the state (x^i,N^i), $i=1,2,\ldots,n$, of the community increases with the negativeness of $\mathcal{R}_\xi$. Because $-\mathcal{R}_\xi$ is a measure of *vigor in production* [Ant85], [AV83], the more vigorous communities are harder to estimate, regardless of the *fixed* observations process used to obtain information, all other things being equal.

In the next chapter we develop a stochastic notion of *vigor* appropriate for a large class of Finslerian h−diffusions and use it in Finslerian filtering. The technicalities are based on quadratic dispersion.

Chapter 5

Finslerian Diffusion and Curvature

As in the well-known Riemannian case, one would expect Finslerian diffusion to be closely related to the curvature of the manifold. In the present chapter we establish such a relationship for Finslerian h-development. A major difficulty is that all methods of relating curvature and diffusion involve normal coordinates, which, unfortunately do not exist for all Finsler manifolds. This is because the exponential map from TM_x to M can, in general, develop a singularity at the origin. Because of this, we restrict ourselves to Berwald spaces, a class of Finsler spaces which can be characterized by the existence of normal coordinates at each point $x \in M$ [Dou28].

The main tool used to relate diffusion and curvature on a Riemannian manifold is the Cartan lemma, which provides an expansion of the metric tensor g_{ij} in normal coordinates in terms of the curvature tensor R^i_{jkl} and its covariant derivatives. Because no analogue of the Cartan lemma for Finsler spaces has been known so far, our first task will be to extend the result to Finsler spacs of Berwald type. This is done in Section 5.1. Then, in Section 5.2 we define the quadratic dispersion of a diffusion and establish a stochastic version of the Taylor formula. In Section 5.3 the quadratic dispersion of Finslerian h-development will be related to the curvature through the Onsager-Machlup term $\frac{R}{12}$, where R is the Cartan scalar curvature. (Equal to the Berwald scalar curvature K in the case of Berwald spaces.) The results of the present chapter were first announced by the authors in [AZ97].

In the case of hv-development the relationship with curvature can be studied on the basis of a result established in Chapter 6, which makes it possible to regard hv-development as a Riemannian development (with some additional drift) on $\mathring{T}M$ equipped with a certain Riemannian metric, see Theorem 6.4. Because the curvature of this Riemannian metric can be expressed in terms of the Finslerian curvature tensors, we will be able to reduce the problem to the well-known case of diffusion on a Riemannian manifold. The details will be presented in Chapter 6. In Section 5.4, we show quadratic dispersion enters a *Finsler filtering problem.* In the long Section 5.5 its relationship to *entropy production* is examined.

5.1 Cartan's Lemma in Berwald Spaces

Throughout this chapter we assume that M is a Berwald space with metric function $L(x, y)$. For the definition and the main properties of Berwald spaces, see Section 4.1.

We denote by $\exp_x$ the *exponential map* from a neighbourhood of $0 \in TM_x$ to a neighbourhood of $x \in M$. Namely, for any $x \in M$ and $y \in TM_x$ we consider the geodesic $\eta(x, y; t)$ starting from x with velocity y at $t = 0$, that is,

$$\frac{d^2\eta^i}{dt^2} + \mathbf{F}^i_{jk}\left(\eta, \frac{d\eta}{dt}\right)\frac{d\eta^j}{dt}\frac{d\eta^k}{dt} = 0 \tag{5.1}$$

with

$$\eta(x, y; 0) = x, \qquad \left.\frac{d}{dt}\right|_{t=0} \eta(x, y; t) = y, \tag{5.2}$$

and we put

$$\exp_x y = \eta(x, y, 1),$$

provided $\eta(x, y, t)$ exists for $t = 1$. When M is a Berwald space, $\exp_x$ is a diffeomorphism from a neighbourhood of $0 \in TM_x$ to a neighbourhood of $x \in M$. The inverse map $\exp_x^{-1}$ form neighbourhood of $x \in M$ to a neighbourhood of $0 \in TM_x$ defines the *normal coordinates* at x, which will be denoted by NC_x for brevity.

For the definition of the Berwald connection coefficients G^i_j and G^i_{jk} appearing in the lemma below, see Section 4.1.

Lemma 5.1. *Let M be a Berwald space. Then for any $x \in M$ and any $y \in TM_x$*

(a) $\mathbf{N}^i_j(x, y) = G^i_j(x, y) = 0$ *in* NC_x,

(b) $\mathbf{F}^i_{jk}(x,y) = \dot{G}^i_{jk}(x,y) = 0$ *in* NC_x,

(c) $\delta_i g_{jk}(x,y) = 0$ *in* NC_x,

(d) $\delta_i\dot{\partial}_j f(x,y) = \dot{\partial}_j\delta_i f(x,y)$ *in* NC_x *for any smooth function* $f : \mathring{T}M \to \mathbf{R}$,

(e) $\delta_h\mathbf{F}^i_{jk}(x,y) + \delta_k\mathbf{F}^i_{hj}(x,y) + \delta_j\mathbf{F}^i_{hk}(x,y) = 0$ *in* NC_x.

Proof. In normal coordinates equation (5.1) becomes

$$\frac{d^2\eta^i}{dt^2} = 0 \quad \text{in } \mathrm{NC}_x \tag{5.3}$$

for any geodesic passing through x. For any $y \in TM_x$ we can take a geodesic $\eta(x,y;t)$ such that (5.2) holds. Putting $t = 0$ in (5.2) and (5.3), we find that

$$G^i(x,y) = \mathbf{F}^i_{jk}(x,y)y^jy^k = 0 \quad \text{in } \mathrm{NC}_x$$

for any $y \in TM_x$. Differentiating this with respect to y^j, we immediately obtain (a) and (b), since $\mathbf{N}^i_j = G^i_j = \dot{\partial}_j G^i$ and $\mathbf{F}^i_{jk} = G^i_{jk} = \dot{\partial}_j G^i_k = \dot{\partial}_j\dot{\partial}_k G^i$ in a Berwald space. Next, (c) follows from (b) because of the h-metricity condition $\delta_i g_{jk} = \mathbf{F}^m_{ij}g_{mk} + \mathbf{F}^m_{ik}g_{jm}$, and (d) obviously follows form (a).

To prove (e) we differentiate both sides of the geodesic equation (5.1) with respect to t to get

$$\frac{d^3\eta^i}{dt^3} + \delta_m\mathbf{F}^i_{jk}\left(\eta, \frac{d\eta}{dt}\right)\frac{d\eta^m}{dt}\frac{d\eta^j}{dt}\frac{d\eta^k}{dt} + \dot{\partial}_m\mathbf{F}^i_{jk}\left(\eta, \frac{d\eta}{dt}\right)\frac{d^2\eta^m}{dt^2}\frac{d\eta^j}{dt}\frac{d\eta^k}{dt}$$
$$+ 2\mathbf{F}^i_{jk}\left(\eta, \frac{d\eta}{dt}\right)\frac{d^2\eta^j}{dt^2}\frac{d\eta^k}{dt} = 0.$$

But in normal coordinates (5.3) holds for any geodesic passing through x, so the above reduces to

$$\delta_m\mathbf{F}^i_{jk}\left(\eta, \frac{d\eta}{dt}\right)\frac{d\eta^m}{dt}\frac{d\eta^j}{dt}\frac{d\eta^k}{dt} = 0 \quad \text{in } \mathrm{NC}_x.$$

Taking η to be a geodesic satisfying (5.2), we obtain

$$\delta_m\mathbf{F}^i_{jk}(x,y)y^my^jy^k = 0 \quad \text{in } \mathrm{NC}_x \tag{5.4}$$

for any $y \in TM_x$. Because, by (d) and (b),

$$\dot{\partial}_n\delta_m\mathbf{F}^i_{jk}(x,y) = \delta_m\dot{\partial}_n\mathbf{F}^i_{jk}(x,y) = 0 \quad \text{in } \mathrm{NC}_x$$

for all $y \in TM_x$, it follows that $\delta_m\mathbf{F}^i_{jk}$ is in fact independent of y. Thus, differentiating (5.4) three times with respect to y, we obtain (e). ∎

Now we are in a position to extend the Cartan lemma to Finsler spaces of Berwald type.

Lemma 5.2. *Let M be a Berwald space. Then for any $x \in M$ and $y \in TM_x$*

(a) $g^{hk} g^{ij} \delta_h \delta_k g_{ij} = \frac{2}{3} R$ *in* NC_x,

(b) $\delta_i \mathbf{F}^i = -\frac{2}{3} R$ *in* NC_x.

Here R is the Cartan scalar curvature (equal to the Berwald scalar curvature K in the case of Berwald spaces.) and $\mathbf{F}^i = g^{jk} \mathbf{F}^i_{jk}$.

Proof. Contracting the indices in (e) of Lemma 5.1, we obtain

$$0 = 2 g^{ij} \delta_i \mathbf{F}^k_{kj} + \delta_i \mathbf{F}^i \quad \text{in } \mathrm{NC}_x.$$

Moreover,

$$\begin{aligned} R = K &= g^{hk} g^{ij} K_{hijk} \\ &= g^{hk} g^{ij} \left(\delta_j \mathbf{F}_{khi} - \delta_k \mathbf{F}_{jhi} \right) \\ &= g^{ij} \delta_j \mathbf{F}^k_{ki} - \delta_i \mathbf{F}^i \qquad \text{in } \mathrm{NC}_x. \end{aligned}$$

Adding both sides of these equalities, we obtain (a). Subtracting the second equality twice from the first one gives (b). ∎

Remark 5.1. The Cartan lemma on Riemannian manifolds involves fewer contraction of indices, so the full curvature tensor features there. This is also possible in the case of Berwald spaces, but the proof is more involved and the above is just what is needed for our purposes.

5.2 Quadratic Dispersion

The definition below is motivated by the notion of kinetic energy $U = \frac{1}{2} g_{ij}(x) \frac{dx^i}{dt} \frac{dx^j}{dt}$ of a smooth trajectory x on a Riemannian manifold as used in classical mechanics. The kinetic energy U can also be defined by

$$\begin{aligned} \mathcal{U}_h(x(t)) &= \frac{1}{2} g_{ij}(x(t)) \left[\exp^{-1}_{x(t)} x(t+h) \right]^i \left[\exp^{-1}_{x(t)} x(t+h) \right]^j \\ &= h^2 U + o(h^2). \end{aligned}$$

The latter definition can readily be extended to diffusions.

Let M be a Berwald space and let (x, y) be a diffusion on $\mathring{T}M$ with generator D.

Definition 5.1. The *quadratic dispersion* of (x,y) is defined by

$$\mathcal{U}_h(x(t),y(t)) = \frac{1}{2}g_{ij}(x(t),y(t))E_t\left\{\left[\exp^{-1}_{x(t)} x(t+h)\right]^i\left[\exp^{-1}_{x(t)} x(t+h)\right]^j\right\},$$

for $t,h \geq 0$, where $E_t = E\{\cdot\,|x(t),y(t)\}$ is the conditional expectation given $x(t), y(t)$.

In normal coordinates this expression takes the form

$$\mathcal{U}_h(x(t),y(t)) = \frac{1}{2}g_{ij}(x(t),y(t))E_t\left\{x^i(t+h)x^j(t+h)\right\} \tag{5.5}$$

Proposition 5.3. *Let M be a Berwald space and let (x,y) be a diffusion on $\mathring{T}M$ with generator D and initial conditions $x(0)=x_0$, $y(0)=y_0$. Then*

$$\begin{aligned}\mathcal{U}_h(x_0,y_0) &= \frac{h}{2}g_{ij}(x_0,y_0)D(x^ix^j)\Big|_{\substack{x=x_0\\y=y_0}} \\ &\quad+\frac{h^2}{4}g_{ij}(x_0,y_0)DD(x^ix^j)\Big|_{\substack{x=x_0\\y=y_0}} + o(h^2) \quad \textit{in } \mathrm{NC}_{x_0}.\end{aligned} \tag{5.6}$$

Proof. For any smooth function f we have

$$\begin{aligned}E_tf(x(t+h),y(t+h)) &= f(x(t),y(t)) + \int_t^{t+h} E_tDf(x(s),y(s))ds \\ &= f(x(t),y(t)) + hDf(x(t),y(t)) + o(h)\end{aligned}$$

Applying the same formula to the integrand above, we obtain the following stochastic version of the Taylor formula (note that D here is a second-order operator):

$$\begin{aligned}&E_tf(x(t+h),y(t+h)) \\ &= f(x(t),y(t)) + \int_t^{t+h}\left[Df(x(t),y(t)) + \int_t^s E_tDDf(x(r),y(r))dr\right]ds \\ &= f(x(t),y(t)) + hDf(x(t),y(t)) + \frac{h^2}{2}DDf(x(t),y(t)) + o(h^2).\end{aligned} \tag{5.7}$$

Now, for any fixed $x_0 \in M$ and $y_0 \in TM_{x_0}$ we take

$$f(x,y) = x^ix^j \quad \text{in } \mathrm{NC}_{x_0}.$$

Then, by (5.5) and (5.7), we obtain (5.6). ∎

5.3 Finslerian Development and Curvature

In this section we establish a relationship between curvature and one kind of Finslerian development, namely, the h-development on a Finsler manifold of Berwald type. It is remarkable that this relationship involves the term $\frac{R}{12}$ of Onsager-Machlup type, where R is the Cartan scalar curvature (equal to the Gauss-Berwald scalar curvature K in a Berwald space).

Theorem 5.4. *Let M be a Berwald space and let (x, y, z) be a stochastic h-development with initial condition $(x_0, y_0, z_0) \in OM$. Then*

$$\mathcal{U}_h(x_0, y_0) = h\frac{\dim M}{2} + h^2\frac{R(x_0, y_0)}{12} + o(h^2). \tag{5.8}$$

Proof. In what follows the arguments (x_0, y_0) will be omitted for brevity. By Theorem 3.10 the generator of (x, y) is

$$\Delta^h = \frac{1}{2}g^{ij}\left(\delta_i\delta_j - \mathbf{F}^k_{ij}\delta_k\right),$$

so we can apply Proposition 5.3 with $D = \Delta^h$. First, we evaluate

$$\Delta^h(x^i x^j) = g^{ij} - \frac{1}{2}x^i\mathbf{F}^j - \frac{1}{2}x^j\mathbf{F}^i.$$

The last two terms vanish in NC_{x_0} by Lemma 5.1 (b), so

$$g_{ij}\Delta^h(x^i x^j)\Big|_{\substack{x=x_0\\ y=y_0}} = g_{ij}g^{ij} = \dim M \quad \text{in } \mathrm{NC}_{x_0},$$

which, by (5.6) gives the first term in (5.8). To find the second term we compute

$$\begin{aligned} g_{ij}\Delta^h\Delta^h(x^i x^j)\Big|_{\substack{x=x_0\\ y=y_0}} &= g_{ij}\Delta^h(g^{ij} - x^i\mathbf{F}^j)\Big|_{\substack{x=x_0\\ y=y_0}} \\ &= \frac{1}{2}g_{ij}g^{mn}\delta_m\delta_n g^{ij} - \delta_j\mathbf{F}^j \\ &= -\frac{1}{2}g^{ij}g^{mn}\delta_m\delta_n g_{ij} - \delta_j\mathbf{F}^j \\ &= -\frac{R}{3} + \frac{2R}{3} = \frac{R}{3} \qquad \text{in } \mathrm{NC}_{x_0} \end{aligned}$$

In this computation we discard all terms which vanish in the normal coordinates by virtue of Lemma 5.1 (b) and the fact that $x^i|_{x=x_0} = 0$ in NC_{x_0}, and then we apply Lemma 5.2 (a) and (b). By (5.6) it gives the second term in (5.8).

We observe that even though (5.8) has been proved in normal coordinates, it is in fact independent of the choice of local coordinates. ∎

The interpretation of the coefficient of the h^2–term in the quadratic dispersion formula as the *negative of vigor*, $-\mathcal{V}$, has its roots in [Ant85] for the deterministic case and in [AES87], on filtering, and [AS87b] on entropy-production. Its deep role in filtering any 2-dimensional signal is presented here in detail in the last part of Chapter 4. In the next section, we consider a 2-dimensional example of Finsler filtering and the role of quadratic dispersion.

5.4 Finslerian Filtering and Quadratic Dispersion

Consider a 2-dimensional Volterra-Hamilton system of Berwald type

$$\begin{cases} \frac{dX^i}{dt} &= N^i \\ \frac{dN^i}{dt} &= \lambda_{(i)} N^i - \Gamma^i_{jk}(x) N^j N^k \end{cases} \tag{5.9}$$

where Γ^i_{jk} are the Berwald connection coefficients of the metric

$$\begin{aligned} L^2(x,y) &= [(y^1)^2 + (y^2)^2] \exp[2(p^2+1)\sigma(x) + 2p \arctan \frac{y^1}{y^2}] \\ &= \overline{L}^2 \cdot \exp[2(p^2+1)\sigma(x)]. \end{aligned} \tag{5.10}$$

The extended dynamical equations for our filtering problem are

$$\begin{aligned} \frac{dN^1}{dt} &= \lambda_1 N^1 - 2\Gamma^1_{12} N^1 N^2 - \Gamma^1_{11}(N^1)^2 - \Gamma^1_{22}(N^2)^2 - \delta_1 F N^1 \\ \frac{dN^2}{dt} &= \lambda_2 N^2 - 2\Gamma^2_{21} N^1 N^2 - \Gamma^2_{11}(N^1)^2 - \Gamma^2_{22}(N^2)^2 - \delta_2 F N^2 \\ \frac{dF}{dt} &= \beta F(N^1 + N^2) + \gamma F^2 - \varepsilon F. \end{aligned} \tag{5.11}$$

Here λ_i, δ_i, $\beta, \gamma, \varepsilon$ are positive constants and

$$\begin{aligned} \Gamma^h_{ij} &= -\dot{\partial}_i \dot{\partial}_j B^{hk} \sigma_k \\ B^{hk} &= \tfrac{1}{2} \overline{L}^2 \cdot \overline{g}^{hk} - y^h y^k \end{aligned} \tag{5.12}$$

with $\sigma_k = \partial_k \sigma$. The Minkowski metric tensor is

$$\begin{aligned} \overline{g}^{(y)}_{11} &= \left[1 + \frac{2py^2(y^1 + py^2)}{(y^1)^2 + (y^2)^2}\right] e^{2p \operatorname{Tan}^{-1} \frac{y^1}{y^2}}, \\ \overline{g}^{(y)}_{22} &= \left[1 - \frac{2(y^1 + py^2) \cdot y^1}{(y^1)^2 + (y^2)^2}\right] e^{2p \operatorname{Tan}^{-1} \frac{y^1}{y^2}}, \\ \overline{g}_{12}(y) &= \overline{g}_{21}(y) = \left[1 - \frac{2p(y^2 - py^1)y^1}{(y^1)^2 + (y^2)^2}\right] e^{2p \operatorname{Tan}^{-1} \frac{y^1}{y^2}}. \end{aligned} \tag{5.13}$$

The Berwald coefficients are explicitly, given as

$$\Gamma^i_{ii} = \widetilde{\sigma}_i = -\Gamma^i_{ij} \qquad (i \neq j),$$

$$\Gamma^i_{ij} = \Gamma^i_{ji} = \widetilde{\sigma}_j \qquad (i \neq j),$$

with

$$\widetilde{\sigma}_1 = \partial_1\sigma - (\partial_2\sigma)p,$$

$$\widetilde{\sigma}_2 = \partial_2\sigma + (\partial_1\sigma)p,$$

as in (4.82). We will now *assume σ is linear in x^i*. Therefore, $\partial_i\sigma = \sigma_i$, $i = 1, 2$ are *constants, taken positive* throughout this section. It follows that the vigor, $\mathcal{V}$, vanishes because of Theorem (4.8). However, following the method of Section 4.7 on Stochastic Nelson Mechanics, we use the *addition of noise ansatz*

$$\xi_t : \begin{cases} dX^i_t = N^i_t dt + e^{\widetilde{\sigma}(x)} \cdot d\eta^i_t \\ \\ dN^i_t = [-\Gamma^i_{jk} N^j N^k + \lambda_{(i)} N^i - \delta_{(i)} N^i F]dt + d\eta^i_t \end{cases} \tag{5.14}$$

for the *signal process* and an *observation process* of the form (4.163) with $n = 2$ and $\delta'_1 = \delta'_2 = \beta$, i.e.

$$\begin{aligned} (dY_t \equiv) \quad dF_t &= \psi(N_t, F_t)F_t dt + dW_t \\ h_t = \quad \psi &= -\varepsilon + \gamma F_t + \beta F_t(N^1_t + N^2_t). \end{aligned} \tag{5.15}$$

This is allowed only because the Finsler geodesic equations (4.82) for the metric (5.10) are precisely those of the Riemannian metric

$$L^2_R = [(y^1)^2 + (y^2)^2] \exp[2\widetilde{\sigma}], \tag{5.16}$$

with $\widetilde{\sigma} = \widetilde{\sigma}_1 x^1 + \widetilde{\sigma}_2 x^2$, $\widetilde{\sigma}_1 = \sigma_1 - \sigma_2 p$, $\widetilde{\sigma}_2 = \sigma_2 + \sigma_1 p$. Naturally, the Riemannian scalar curvature vanishes for this metric, so the Riemannian and Finslerian quadratic dispersions agree.

The filtering problem is solved by the expression (4.164) for the Zakai equation (4.130) with $n = 2$ and $\frac{1}{12}\, \mathrm{IR}_\xi = 0$. Also, $\pi_t(f)$ given by (4.165). However, the h–diffusion for this filtering problem can be chosen Riemannian *because* the Berwald connection coefficients are constants. To see that this is true, suppose there exists a scalar function $\psi(x)$ so that

$$\partial_1\psi = \frac{\sigma_1 - p\sigma_2}{1 + p^2}, \quad \partial_2\psi = \frac{\sigma_2 + p\sigma_1}{1 + p^2}. \tag{5.17}$$

Integrability of (5.17) implies

$$\Delta\sigma \equiv \partial_1\partial_1\sigma + \partial_2\sigma_2\sigma = 0, \tag{5.18}$$

so that $\sigma(x)$ must be *harmonic*. Therefore, the Riemannian metric, $\widetilde{L}^{\,2} = [(y^1)^2 + (y^2)^2]e^{2\psi(x)}$, whose geodesics are

$$\begin{aligned} \tfrac{dy^1}{ds} + f(x)(y^1)^2 + 2g(x)y^1y^2 - f(x)(y^2)^2 &= 0 \\ \tfrac{dy^2}{ds} - g(x)(y^1)^2 + 2f(x)y^1y^2 + g(x)(y^2)^2 &= 0 \end{aligned} \tag{5.19}$$

with

$$f(x) = \frac{\sigma_1 - p\sigma_2}{1+p^2}, \qquad g(x) = \frac{\sigma_2 + p\sigma_1}{1+p^2},$$

must have vanishing curvature. This follows directly from Theorem (4.8) with $p = 0$ and $\psi(x)$ replacing $\sigma(x)$.

Therefore, if $\sigma(x)$ is taken to be a quadratic polynomial, for example, the above trick can not be used to solve the filtering problem. In the same way, filtering the myxomatosis model dynamics which involves the mth–root metric will not yield to this method, except for the $m = 2$ case, [AS87a], [AS88a]. Both these filtering problems, then, are open at the time of this writing.

5.5 Entropy Production and Quadratic Dispersion

Let M be an orientable n–dimensional C^∞–manifold without boundary (e.g., $\mathbb{R}^n$). Associated to a Riemannian geometry g_{ij} on M, is the *Riemannian volume form* (or invariant measure), $d\mu$, defined by

$$d\mu(x) = \sqrt{g}\,(x)dx \tag{5.20}$$

where $\sqrt{g}\,(x) = \left(\det\left(g_{ij}(x)\right)\right)^{1/2}$ in local co-ordinates $x = \{x^i\}$, and where dx denotes the standard Lebesgue measure on $\mathbb{R}^n$.

The function $\sqrt{g}\,(x)$ does not define an invariant scalar function on M. In fact, its transformation law is

$$\sqrt{g}\,(\widetilde{x}) = \sqrt{g}\,(x)|J(x,\widetilde{x})|, \tag{5.21}$$

where $|J(x,\widetilde{x})|$ is the determinant of the Jacobian matrix of the transformation $x \mapsto \widetilde{x}$ of local co-ordinates. However, if we are given a *second*

Riemannian geometry g'_{ij} (defining the same orientation as g_{ij}), then it is clear from the universal relation (5.21) that the *ratio*

$$\phi = \sqrt{g}\,/\sqrt{g'}\,, \tag{5.22}$$

does define a C^∞ invariant scalar function on M. Furthermore, ϕ is strictly positive everywhere.

Let $\langle u, v\rangle_{g'}$ denote the L_2–inner products defined by the invariant measures $d\mu, d\mu'$ for the geometries g_{ij} and g'_{ij}, respectively (5.20). Here, u and v are suitable scalar functions on M which are square integrable with respect to *both* measures (e.g., C^∞ functions with compact support, or which decay to zero rapidly enough at infinity).

If $\mathbb{L} : C^r(M) \to C^{r-k}(M)$ is a linear operator on C^r–scalar functions on M $(r \geq k \geq 0)$, denote by $\mathbb{L}^+$, respectively L^*, the L_2–adjoints of $\mathbb{L}$ with respect to $d\mu$ and $d\mu'$; i.e.,

$$\langle \mathbb{L}u; v\rangle_g = \langle u; \mathbb{L}^+ v\rangle_g \quad \text{and} \quad \langle \mathbb{L}u; v\rangle_{g'} = \langle u; \mathbb{L}^* v\rangle_{g'} \tag{5.23}$$

for any pair u, v for which both inner products exist. We then have

$$\mathbb{L}^*(\phi u) = \phi L^+(u) \tag{5.24}$$

for any $u \in C^r(M)$, where ϕ is given by (5.22).

To establish (5.24), let $v \in C^\infty(M)$ have compact support. Then the inner products in (5.23) exist for any $u \in C^r(M)$. From (5.20) and (5.22), we have that $d\mu = \phi d\mu'$, whence,

$$\langle v; u\rangle_g = \langle \phi v; u\rangle_{g'} = \langle v; \phi u\rangle_{g'}. \tag{5.25}$$

Thus

$$\begin{aligned}
\langle v; \mathbb{L}^*(\phi u)\rangle_{g'} &= \langle \mathbb{L}v; \phi u\rangle_{g'} && \text{by (5.23)}\\
&= \langle \mathbb{L}v; u\rangle_g && \text{by (5.25)}\\
&= \langle v; \mathbb{L}^+ u\rangle_g && \text{by (5.23)}\\
&= \langle v; \phi \mathbb{L}^+ u\rangle_{g'} && \text{by (5.25).}
\end{aligned}$$

Since v is arbitrary, (5.24) now follows from the non-degeneracy of the L_2–inner product.

We apply (5.24) to the case $L = \widetilde{\mathbb{D}} = \mathbb{D} + V$, where V is a continuous potential function on M, and $\mathbb{D}$ is a diffusion operator of the form,

$$\mathbb{D} = \frac{1}{2}\,\Delta_g + \boldsymbol{h}. \tag{5.24a}$$

Associated to such an operator are *two* forward diffusion equations

$$\frac{\partial v}{\partial t} = \widetilde{\mathbb{D}}^{+} v \qquad \text{and} \qquad \frac{\partial u}{\partial t} = \widetilde{D}^{*} u, \tag{5.26}$$

where u, v are scalar functions on $M \times (0, \infty)$ which are C^2 in the space variables and C^1 in the time variable. The relation (5.24) allows us to establish a relation between solutions of the two equations (5.26). Thus, we have

Proposition 5.5. *Let v be a solution of the first equation* (5.26) *with initial distribution $v_0 = \lim_{t\downarrow 0} v_t$. Then $u = \phi v$ is a solution of the second equation* (5.26) *with initial distribution $u_0 = \phi v_0 = \lim_{t\downarrow 0} u_t$.*

To use Proposition 5.5, we need to compute the adjoint $\widetilde{\mathbb{D}}^{+}$ in terms of invariants involving the g_{ij}–geometry. To this end, we recall the definition of the *Riemannian divergence* of a contravariant vector field X;

$$\operatorname{div}_g(X) = \frac{1}{\sqrt{g}}\, \partial_i(\sqrt{g}\, X^i), \tag{5.27}$$

in any local co-ordinate system.

Now recall (5.24a) that $\widetilde{\mathbb{D}} = \frac{1}{2}\,\Delta_g + \boldsymbol{h} + V$, where Δ_g is the invariant Laplacian for the g_{ij}–geometry, and $\boldsymbol{h}$ is a contravariant vector field on M. We may regard $\boldsymbol{h}$ as a first order differential operator on $C^r(M)$ $(r \geq 1)$; $\boldsymbol{h} = h^i \partial_i$. We then have

$$\boldsymbol{h}^{+} = -\boldsymbol{h} - \operatorname{div}_g \boldsymbol{h}. \tag{5.28}$$

To establish (5.28), we have that, for suitable test functions u and v,

$$\begin{aligned} \langle \boldsymbol{h}(u); v\rangle_g \;&+\; \langle u; \boldsymbol{h}(v) + v \operatorname{div}_g \boldsymbol{h}\rangle_g \\ &= \int_M \{\boldsymbol{h}(u)v + u\boldsymbol{h}(v) + uv \operatorname{div}_g \boldsymbol{h}\}\, d\mu \\ &= \int_M \operatorname{div}_g(uv\boldsymbol{h})\, d\mu \quad \text{using (5.27)}. \end{aligned}$$

Supposing u to have compact support, this integral is zero by the divergence theorem and (5.28) now follows.

Recalling that Δ_g is self-adjoint in the g_{ij}–geometry, (5.28) now shows that

$$\text{If} \quad \widetilde{\mathbb{D}} = \frac{1}{2}\,\Delta_g + \boldsymbol{h} + V, \quad \text{then} \quad \widetilde{\mathbb{D}}^{+} = \frac{1}{2}\,\Delta_g - \boldsymbol{h} - \operatorname{div}_g \boldsymbol{h} + V. \tag{5.29}$$

Note that (5.29) implies that $\widetilde{\mathbb{D}}$ is self-adjoint if and only if $\boldsymbol{h} = 0$.

From (5.25), taking u to be the constant function 1, we obtain, $\widetilde{\mathbb{D}}^*(\phi) = \phi \cdot \widetilde{\mathbb{D}}(1)$, and from (5.29), $\widetilde{\mathbb{D}}^+(1) = V - \operatorname{div}_g \boldsymbol{h}$. Thus, we have shown

Proposition 5.6. *Let ϕ be as in (5.22), and suppose that $V - \operatorname{div}_g \boldsymbol{h} = 0$. Then $\widetilde{\mathbb{D}}^*(\phi) = 0$; i.e., ϕ is a C^∞*–stationary *density for the forward diffusion operator $\widetilde{\mathbb{D}}^* = \mathbb{D}^* + V$.*

Note in particular, that Proposition 5.6 holds if $V = \operatorname{div}_g \boldsymbol{h} = 0$. As is well known, the requirement that $\boldsymbol{h}$ be divergence free means that $\boldsymbol{h}$ preserves volumes in the g_{ij}–geometry. The more general statement of Proposition 5.6 can now be interpreted as stating that a sufficient condition for the existence of a stationary density is that any tendency of the (covariant) drift vector $\boldsymbol{h}$ to compress or expand volumes, should be compensated by the existence of a balancing gradient field.

We resume the task of computing $\widetilde{\mathbb{D}}^*$ in terms of invariants of the g_{ij}–geometry. From (5.28), we have

$$\widetilde{\mathbb{D}}^* = \frac{1}{2}\Delta_g^* - \boldsymbol{h} - \operatorname{div}_{g'} \boldsymbol{h} + V. \tag{5.30}$$

To compute $\operatorname{div}_{g'} \boldsymbol{h}$, we note the formula

$$\operatorname{div}_g(\psi X) = \psi \operatorname{div}_g X + X(\psi), \tag{5.31}$$

where X is a C^1 contravariant vector field on M (regarded as a first order differential operator), and ψ is a C^1–scalar function. Using the fact that $\sqrt{g} = \phi\sqrt{g'}$ (5.22), (5.27), and (5.31), we find that

$$\begin{aligned} \operatorname{div}_{g'}(\boldsymbol{h}) &= \phi \operatorname{div}_g\left(\frac{1}{\phi}\boldsymbol{h}\right) \\ &= \operatorname{div}_g(\boldsymbol{h}) - \boldsymbol{h}(\ell n\, \phi). \end{aligned} \tag{5.32}$$

Now let $(X;Y)_g$ denote the *Riemannian inner product* for vector fields X and Y, i.e.,

$$(X;Y)_g = g_{ij}X^iY^j, \tag{5.33}$$

and let $|X|_g = (X:X)_g^{1/2}$, denote the associate *Riemannian norm.* Finally, for ψ a C^1–scalar function, let $\operatorname{grad}_g\psi$ denote the *Riemannian gradient* of ψ,

$$\operatorname{grad}_g^i\psi = g^{ij}\partial_j\psi. \tag{5.34}$$

We may write the second term of (5.31) as

$$X(\psi) = (X; \operatorname{grad}_g \psi)_g. \tag{5.35}$$

From (5.30), we must compute Δ_g^*. We have

$$\Delta_g^* = \Delta_g - 2 \operatorname{grad}_g(\ell n\ \phi) + \mathcal{L}_g(\phi), \tag{5.36}$$

where the vector field $\operatorname{grad}_g(\ell n\ \phi)$ is regarded as a first order differential operator, and $\mathcal{L}_g(\phi)$ is the *invariant scalar function*

$$\mathcal{L}_g(\phi) = |\operatorname{grad}_g(\ell n\ \phi)|_g^2 - \Delta_g(\ell n\ \phi). \tag{5.37}$$

To establish (5.36), recall that $\Delta_g(u) = \operatorname{div}_g \operatorname{grad}_g(u)$. Thus

$$\begin{aligned} \Delta_g^*(u) &= \phi \Delta_g^+(\tfrac{1}{\phi}\, u) \qquad \text{(from (5.24))} \\ &= \phi \Delta_g(\tfrac{1}{\phi}\, u) \end{aligned}$$

since Δ_g is self-adjoint in the g_{ij}–geometry.

Now, use of (5.31), (5.35) and the relation $\operatorname{grad}_g(\psi u) = \psi \operatorname{grad}_g u + u \operatorname{grad}_g \psi$, shows that, for any C^2–scalar function ψ,

$$\Delta_g(\psi u) = \psi \Delta_g u + 2(\operatorname{grad}_g \psi; \operatorname{grad}_g u) + u \Delta_g \psi.$$

Applying this with $\psi = 1/\phi$, and noting that $(1/\phi) \operatorname{grad}_g \phi = \operatorname{grad}_g(\ell n\ \phi)$, we find that

$$\Delta_g^*(u) = \Delta_g u - 2(\operatorname{grad}_g(\ell n\, \phi); \operatorname{grad}_g u)_g + \{2|\operatorname{grad}_g(\ell n\ \phi)|_g^2 - \frac{1}{\phi}\, \Delta_g \phi\} u.$$

Finally, we note, using (5.31) and (5.35), that

$$\Delta_g(\ell n\ \phi) = \frac{1}{\phi}\, \Delta_g \phi - |\operatorname{grad}_g(\ell n\ \phi)^2|_g.$$

(5.36) and (5.37) now follow.

To summarize, from (5.30), (5.32), and (5.36), we have shown

Proposition 5.7. *Let* $\mathbb{D} = \frac{1}{2}\Delta_g + \boldsymbol{h}$. *Then*

$$\mathbb{D}^* = \frac{1}{2}\Delta_g - (\boldsymbol{h} + \operatorname{grad}_g(\ell n\, \phi)) + \{\frac{1}{2}\, \mathcal{L}_g(\phi) + \boldsymbol{h}(\ell n\, \phi) - \operatorname{div}_g(\boldsymbol{h})\},$$

where ϕ *is given by* (5.22), $\mathcal{L}_g(\phi)$ *by* (5.37), *and* $\mathbb{D}^*$ *is the adjoint of* $\mathbb{D}$ *with respect to the metric* g'_{ij}.

Another useful expression for $\mathbb{D}^*$ is now easily derived from Proposition 5.7, namely

$$\mathbb{D}^*(u) = \operatorname{div}_{g'} \{\frac{1}{2}\, \phi \operatorname{grad}_g(\phi^{-1}u) - u\boldsymbol{h}\}. \tag{5.38}$$

To see this, express $\operatorname{div}_{g'}$ in terms of div_g using (5.32), then expand the right-hand side of (5.38) using (5.31) and (5.35), and compare the resulting expression with that given in Proposition 5.7.

Finally, it is worth noting that nothing in the discussion of this section depends on any properties of the metric g'_{ij} other than its volume from $d\mu'$. It therefore suffices just to postulate the existence of such a volume form (defining the same orientation as $d\mu$), rather than a full Riemannian geometry. The essential point is that the forward diffusion equation has $\mathbb{D}^*$ as above, if the Chapman-Kolmogorov equations for the process $\{X_t\}$ are defined with respect to the measure $d\mu'$.

We return to the case $M = \mathbb{R}^n$, and take g'_{ij} to be the standard Euclidean metric. Thus, for a diffusion operator $\mathbb{D}$ given by (5.24a), $\mathbb{D}^*$ *refers to the formal Euclidean adjoint.*

Given an initial distribution u_0 on $\mathbb{R}^n$, we consider solutions $u(x,t)$ to the forward diffusion equation $(\widetilde{\mathbb{D}}^* - (\partial/\partial t))u = 0$ with $\lim_{t\downarrow 0} u(x,t) = u_0(x)$. (Here, as above, $\widetilde{\mathbb{D}}^* = \mathbb{D}^* + V$ for some potential function V.) We suppose $u(x,t)$ is C^2 in x and C^1 in t for $t > 0$. Under certain *additional* boundedness and smoothness conditions, an explicit expression for $u(x,t)$ can be given in terms of a functional integral; e.g., [Kun83], part I, Sect. 5. *However, we will not make use of such expressions here, and we merely suppose that such solutions exist and are sufficiently well-behaved.*

Generally, we are interested in non-negative solutions *which remain positive somewhere* (and hence on some non-empty open set) *throughout their evolution.* Thus, we rule out solutions which decay to zero after a finite time. In order to normalize such a solution to form a *probability density,* we must find conditions under which

$$N_t = \int_{\mathbb{R}^n} u(x,t)dx \tag{5.39}$$

is finite. Note that our assumptions on the positivity of $u(x,t)$ imply that $N_t > 0$ for all $t > 0$.

We shall consider the finiteness of (5.39) by making assumptions concerning the Riemannian geometry g_{ij} associated to the operator $\mathbb{D}$. To do this, it is convenient to use Proposition 5.5 to express the solution $u(x,t)$ in the form

$$u(x,t) = v(x,t)\phi(x), \tag{5.40}$$

where $v(x,t)$ is a solution of $(\mathbb{D}^+ - (\partial/\partial t))(v) = 0$, with $\lim_{t\downarrow 0} v(x,t) = \phi^{-1}(x)u_0(x)$. Here, $\phi(x) = \sqrt{g}\,(x)$ is given by (5.22) with g'_{ij} the standard Euclidean metric (so that $g'_{ij}(x) = \delta_{ij}$ in standard co-ordinates). Since ϕ is strictly positive everywhere, the assumptions concerning the positivity of $u(x,t)$ are equivalent to similar assumptions for $v(x,t)$.

The reason for preferring (5.40) is that *we shall assume that* $\phi(x)$ *has strong convergence properties.* In particular, we want to assume that $\mathbb{R}^n$ has *finite volume* in the g_{ij} geometry. But we shall require rather more than this. To state what is needed to justify the formal manipulations which follow, we first define *two* notions of *domination* for (Lebesgue measurable) functions on $\mathbb{R}^n$, both of which imply that such functions are integrable with respect to the measure $d\mu(x) = \phi(x)dx$, given suitable assumptions on the g_{ij} metric.

First, we shall say that a function $f(x)$ is *polynomially dominated* (p.d.) if

$$|f(x)| \leq C(1 + |x|^k) \tag{5.41}$$

for some positive constant C, and non-negative integer k.

The notion of p.d. is independent of the metric g_{ij}, but we shall also need a (weaker) notion of domination which is geared to the specific metric we are considering. Thus, we shall say that $f(x)$ is *allowably dominated* (a.d.) if

$$|f(x)| \leq C(1 + |x|^k)\phi^{-\alpha}(x) \tag{5.42}$$

with C and k as in (5.41) and $0 \leq \alpha < 1$. The idea is that $|f(x)|$ should be allowed to grow very fast as $|x| \to \infty$, but not quite as fast as $\phi^{-1}(x)$.

In terms of these notions of domination, we shall make specific assumptions concerning the metric g_{ij} and the solutions $v(x,t)$. Thus, for the metric, we assume

(i) Each $g_{ij}(x)$ is p.d.

(ii) $g^{ij}(x)$ and its first partial derivatives are a.d.

(iii) $\ell n\ \phi$ and its first and second order partial derivatives are p.d. (5.43)

(iv) $\phi(x) \leq e^{-K|x|^\sigma}$ some positive constants K and σ, and for all sufficiently large $|x|$.

Note that (iv) implies that $\phi(x) \to 0$ as $|x| \uparrow \infty$. Also, (i) implies that the $g_{ij}(x)$ are fairly tame at infinity, while (ii) implies that $g^{ij}(x)$ can be much wilder as $|x| \uparrow \infty$ (but not as wild as ϕ^{-1}). From (iv), we have $\phi^{-\alpha}(x) \geq 1$ for all positive α and sufficiently large $|x|$. Hence, (5.42) implies (5.41), so that a.d. is weaker than p.d.

We collect together the technical properties we shall use, which follow from the assumptions (5.43).

Lemma 5.8. (i) *If* $f(x)$ *is* a.d. *then* $f \in L^1(\mathbb{R}^n; d\mu)$.

(ii) *If* $f(x)$ *is* p.d. *and* $g(x)$ *is* a.d., *then* $f(x)g(x)$ *is* a.d.

(iii) *If* $f(x)$ *and its first partial derivatives are* p.d., *then* $|\text{grad}_g f|$ *is* a.d.

(iv) *If* X *is a vector field on* $\mathbb{R}^n$ *such that* $|X|$ *is* a.d., *and* $f(x)$ *is as in* (iii), *then* $X(f)$ *is* a.d.

(v) *If* $|X|$ *and* div X *are* a.d., *then so is* $\text{div}_g X$.

(vi) *If* $f(x)$ *and its first and second order partial derivatives are* p.d., *then* $\Delta_g f$ *is* a.d.

Proof. (i) follows from (5.42), (5.43)(iv) and the fact that

$$\int_{\mathbb{R}^n} |X|^k e^{-K(1-\alpha)|X|^\sigma} dx < \infty$$

for σ, $K(1-\alpha) > 0$; and k a non-negative integer.

(ii) is obvious.

(iii) First observe that a vector field X has the property that $|X|$ is a.d. if and only if each component X^i is a.d. Now $\text{grad}_g^i f$ is a sum of a.d. functions by (5.43)(ii) and hence is a.d.

(iv) $|X(f)| = |(X; \text{grad } f)| \leq |X| \cdot |\text{grad } f|$, and this is the product of an a.d. function and a p.d function, so the result follows by Lemma 5.8(ii).

(v) $|\text{div}_g X| = |\text{div } X + X(\ell\text{n } \phi)| \leq |\text{div } X| + |X(\ell\text{n } \phi)|$, by (5.32).
The first term is a.d. by hypothesis, while the second term is a.d. by (5.43)(iii) and Lemma 5.8(iv).

(vi) $\Delta_g f = \text{div}_g(\text{grad}_g f) = \text{div grad}_g f + (\text{grad}_g f)(\ell\text{n } \phi)$ by (5.32). The second term is a.d by (5.43)(iii) and Lemma 5.8(iii).

For the first term, we have

$$\operatorname{div} \operatorname{grad}_g f = \partial_i(g^{ij}\partial_j f) = g^{ij}\partial_i\partial_j f + \partial_i(g^{ij})\partial_j f.$$

The result now follows from (5.43)(ii) and the fact that $\partial_j f$ and $\partial_i\partial_j f$ are p.d., and Lemma 5.8(ii). ∎

For the solution $v(x,t)$ of (5.40) we shall assume that $v(x,t)$ *and its first and second order (space) partial derivatives are* p.d. Note, in particular, that this implies that $|\operatorname{grad}_g v_t|$ and $\Delta_g v_t$ are a.d. (by Lemma 5.8(iii) and (vi)). We shall also suppose that $|\boldsymbol{h}|$ *and* $\operatorname{div} \boldsymbol{h}$ *are* a.d., *and that* V *is measurable and* a.d. Note that this implies that $\boldsymbol{h}(v_t)$ and $(V - \operatorname{div}_g \boldsymbol{h})v_t$ are a.d. (by Lemma 5.8(ii) and (iv)). It now follows, from (5.29), *that* $\widetilde{\mathbb{D}}^+ v_t = \partial v_t/\partial t$ *is* a.d., *and hence is* L' *with respect to* $d\mu$ (Lemma 5.8(i)).

It follows from (5.40) and the preceding discussion that $0 < N_t < \infty$. Thus, the *probability density*

$$p(x,t) = u(x,t)/N_t \tag{5.44}$$

exists. Note, however, that $p(x,t)$ does *not* in general satisfy the forward diffusion equation $(\widetilde{\mathbb{D}}^* - \partial/\partial t)p = 0$. In fact, assuming that N_t is differentiable (which will be justified below), we have

$$\frac{\partial p}{\partial t} = \widetilde{\mathbb{D}}^*(p) - p\,\frac{d}{dt}\,(\ell\text{n } N_t). \tag{5.45}$$

The additional term on the right-hand side of (5.45) can be obtained by replacing V by the (time-dependent) potential $V - (d/dt)(\ell\text{n } N_t)$, and so it acts as an additional force acting to counterbalance any loss or gain of "total mass" in time. Note that, if *total biomass is conserved in time;* i.e., if $N_t = N_0$ for all $t \geq 0$, then (5.45) reduces to the original diffusion equation, and almost all properties of $u(x,t)$ are inherited by $p(x,t)$.

To see what is required for "conservation of mass" recall that our assumptions imply that $\partial v(x,t)/\partial t = \widetilde{\mathbb{D}}^+(v(x,t))$ is a.d. Hence, by Lemma 5.8(i), the differentiation under the integral sign is justified in the computation

$$\begin{aligned}\frac{dN_t}{dt} &= \int_{\mathbb{R}^n} \frac{\partial v(x,t)}{\partial t}\, d\mu(x)\\ &= \int_{\mathbb{R}^n} \mathbb{D}^+(v(x,t))d\mu(x) + \langle V\rangle_t\\ &= \int_{\mathbb{R}^n} \mathbb{D}^*(u(x,t))dx + \langle V\rangle_t.\end{aligned} \tag{5.46}$$

Here $\langle f\rangle_t$ denotes the *expected value* of the function $f(x)$ with respect to the measure $v(x,t)d\mu(x) = u(x,t)dx$. Note that, our assumptions on V and $\boldsymbol{h}$ imply that both $\mathbb{D}^+(v(x,t))$ and $V(x)v(x,t)$ are a.d., so that both right-hand integrals exist. Of course, (5.46) also shows that N_t is differentiable, and so justifies (5.45).

We would now like to apply (5.38), and conclude from the (Euclidean) divergence theorem, that the first integral on the right-hand side of (5.46) vanishes. Clearly, we must have fairly stringent boundary conditions "at infinity" for this to be justified. Assuming such conditions for the moment, (5.46) then integrates to give

$$N_t = N_0 + \int_0^t \langle V\rangle_s ds. \tag{5.47}$$

This shows that *the accumulation or loss in time is due entirely to the potential* V (provided there are no sources or sinks "at infinity"). In particular, *if* $V=0$, *then* $N_t = N_0$ *for all* t, *and we have "conservation of total biomass."* Another notable case is when $V=E$, a non-zero constant, and we then have $N_t = N_0 e^{Et}$, so that total biomass grows or decays exponentially.

To see what conditions are required for (5.47) to hold, we apply the divergence theorem to (5.38) and use (5.40) to obtain

$$\begin{aligned}\int_{\mathbb{R}^n} \mathbb{D}^*(u(x,t))dx &= \frac{1}{2}\lim_{R\uparrow\infty}\int_{S_R}(\boldsymbol{v};\ \phi\,\mathrm{grad}_g(v_t))dA_R \\ &\quad -\lim_{R\uparrow\infty}\int_{S_R}(\boldsymbol{v};\phi v_t\boldsymbol{h})dA_R,\end{aligned} \tag{5.48}$$

where S_R is the (Euclidean) sphere of radius R, centre 0, in $\mathbb{R}^n$, dA_R the induced Euclidean measure on S_R, and $\boldsymbol{v}$ the outward pointing unit normal.

Now note that, for any vector field X on $\mathbb{R}^n$, we have

$$\left|\int_{S_R}(\boldsymbol{v};\phi X)dA_R\right| \le V(S)R^{n-1}\sup_{v\in S}\{|X(R\boldsymbol{v})|\phi(R\boldsymbol{v})\},$$

where S is the unit sphere in $\mathbb{R}^n$, and $V(S)$ its Euclidean volume. Thus,

$$\int_{\mathbb{R}^n}\mathrm{div}\,(\phi X)dx = 0 \quad \text{if} \quad \phi(x)|X|^{n-1}|X(x)| \to 0 \text{ as } |X|\uparrow\infty. \tag{5.49}$$

In particular, *this is true if* $|X(x)|$ *is* a.d. (see (5.42) and (5.43)(iv)).

From (5.48) and (5.49), we conclude that (5.46) holds if $|v_t\boldsymbol{h}|$ and $|\mathrm{grad}_g v_t|$ are a.d. But this is implied by our assumptions on g^{ij}, v_t, and $\boldsymbol{h}$. Hence, *with our stated assumptions*, (5. 47) *holds.* ∎

We shall retain our previous assumptions, but now also suppose that $V = 0$, so that $N_t = N_0$ for all t, and the probability density (5.44) is a stationary solution of the forward diffusion equation (5.45). Consider the (information) *entropy*

$$\begin{aligned} S_t &= -\int_{\mathbb{R}^n} \ell\text{n}\,(p(x,t))p(x,t)dx \\ &= -\int_{\mathbb{R}^n} \ell\text{n}\,(p(x,t))q(x,t)d\mu(x), \end{aligned} \tag{5.50}$$

where $q(x,t) = v(x,t)/N_0$, and $p(x,t) = \phi(x)q(x,t)$.

It is usual to take $(p\,\ell\text{n}\,p)(x,t) = 0$ if $p(x,t) = 0$ for some (x,t). However, if we allow p_t to have *finite* zeros, certain additional complications arise in the arguments which follow. Although this can to some extent be overcome by additional assumptions, it seems best, to retain reasonable simplicity and avoid too many side issues, to *assume that p_t is strictly positive on $\mathbb{R}^n$ for all $t > 0$.*

Again, to justify the formal manipulations which follow, we shall need to assume not only that q_t, but also that $\ell\text{n}\ q_t$ is reasonably well behaved "at infinity". Specifically, *we must ensure that $|\ell\text{n}\,q_t|$ does not grow "too fast" if $q_t \to 0$ in some direction.* The easiest way to ensure this is to *suppose that $\ell\text{n}\ q_t$ and its first and second order (space) partial derivatives are* p.d.

We also note the following useful fact

$$\textit{If} \quad f(x) \quad \text{is a.d., } \textit{then so are} \quad f\,\ell\text{n}\,\phi \quad \textit{and} \quad f\,\ell\text{n}\,|f|. \tag{5.51}$$

That $f\ \ell\text{n}\ \phi$ is a.d. follows from (5.43)(iii) and Lemma 5.8(ii). For $f\ \ell\text{n}\,|f|$, we note first that (5.42) and (5.43)(iii), Lemma 5.8(ii) imply that

$$(|f|\,\ell\text{n}\,|f|)(x) \le D(1 + |x|^\ell)\phi^{-p}.$$

But, from the fact that $y\ \ell\text{n}\ y \ge -1/e$ for any $y \ge 0$, we have that $|f|\,\ell\text{n}\,|f|$ is bounded below, and so (5.51) follows.

The existence of the integral (5.50) follows from the estimate

$$|q_t\,\ell\text{n}\ p_t| = |q_t\,\ell\text{n}\ q_t + q_t\,\ell\text{n}\ \phi| \le |q_t\,\ell\text{n}\ q_t| + |q_t\,\ell\text{n}\ \phi|,$$

and the fact that all these functions are assumed p.d. (and hence a.d.).

We now wish to compute dS_t/dt. Differentiating (5.50) formally, we obtain

$$\frac{dS_t}{dt} = -\int_{\mathbb{R}^n} \left(\frac{\partial q_t}{\partial t}\,\ell\text{n}\ q_t + \frac{\partial q_t}{\partial t}\,\ell\text{n}\ \phi + \frac{\partial q_t}{\partial t}\right)d\mu. \tag{5.52}$$

To justify the differentiation under the integral sign, it is enough to show that each of the three terms in the integrand are a.d. Now, as noted previously, our assumptions imply that $\mathbb{D}^+(q_t) = \partial q_t/\partial t$ is a.d. Hence, $(\partial q_t/\partial t)\,\ell\text{n}\,\phi$ is a.d. by (5.51), while $(\partial q_t/\partial t)\,\ell\text{n}\,q_t$ is a.d. by Lemma 5.8(ii) and our assumption that $\ell\text{n}\,q_t$ is p.d.

We proceed with the formal analysis of (5.52), Noting that we have shown previously that

$$\int_{\mathbb{R}^n} \frac{\partial q_t}{\partial t}\, d\mu = \int_{\mathbb{R}^n} \mathbb{D}^*(p(x,t))dx = 0,$$

(5.52) reduces to

$$\frac{dS_t}{dt} = -\int_{\mathbb{R}^n} \frac{\partial p_t}{\partial t}\,\ell\text{n}\; p_t \cdot dx = -\int_{\mathbb{R}^n} \mathbb{D}^*(p_t)\,\ell\text{n}\,p_t \cdot dx. \tag{5.53}$$

To compute the latter integral, we first use (5.38) to obtain the formula

$$u\mathbb{D}^*(\phi v) - \phi v\mathbb{D}(u) = \text{div}\,\big[\phi\{\tfrac{1}{2}(u\;\text{grad}_g v - v\;\text{grad}_g u) - w\boldsymbol{h}\}\big]. \tag{5.54}$$

Thus, it follows from (5.49) that *is* $|\frac{1}{2}\,(u\;\text{grad}_g v - v\;\text{grad}_g u) - uv\boldsymbol{h}]$ *is* a.d., *then*

$$\int_{\mathbb{R}^n} u\mathbb{D}^*(\phi v)dx = \int_{\mathbb{R}^n} v\mathbb{D}(u)d\mu. \tag{5.55}$$

We wish to apply the above with $u = \ell\text{n}\;p_t = \ell\text{n}\;q_t + \ell\text{n}\;\phi$, and $v = q_t$. The vector field inside the $\{\quad\}$ in (5.54) then becomes

$$\begin{aligned}\frac{1}{2}\,\{\text{grad}_g(q_t)\ell\text{n}\;q_t \;&+\; \text{grad}_g(q_t)\ell\text{n}\;\phi - q_t\,\text{grad}_g(\ell\text{n}\;q_t) - q_t\,\text{grad}_g(\ell\text{n}\;\phi)\}\\ &-\; (q_t\;\ell\text{n}\;q_t + q_t\;\ell\text{n}\;\phi)\boldsymbol{h}.\end{aligned} \tag{5.56}$$

Recalling that we are assuming that q_t, $\ell\text{n}\;q_t$ and $\ell\text{n}\;\phi$ are p.d. and that $|\boldsymbol{h}|$ is a.d., it follows from Lemma 5.8(ii), that $|(q_t\,\ell\text{n}\,q_t + q_t\,\ell\text{n}\;\phi)\boldsymbol{h}|$ is a.d. Again, (5.43)(iii) together with Lemma 5.8(ii) and (iii) imply that $q_t|\text{grad}_g(\ell\text{n}\;\phi)|$ is a.d. Also, $q_t\,\text{grad}_g(\ell\text{n}\,q_t) = \text{grad}_g q_t$ is a.d. by Lemma 5.8(iii), while $|\text{grad}_g(q_t)\ell\text{n}\,\phi|$ is a.d. by Lemma 5.8(iii) and (5.43)(iii). Finally, $|\text{grad}_g(q_t)\ell\text{n}\;q_t|$ is a.d. by the Lemma 5.8(iii) and the assumption that $\ell\text{n}\;q_t$ is p.d. We conclude that (5.56) is a.d., and we have therefore obtained the relation (5.55) with $u = \ell\text{n}\;p_t$ and $v = q_t$. Hence (5.53) reduces to

$$\frac{dS_t}{dt} = -\int_{\mathbb{R}^n} p_t\mathbb{D}(\ell\text{n}\;p_t)dx. \tag{5.57}$$

Now note, from (5.24a) that

$$\begin{aligned} \mathbb{D}(\ell n\, p) &= \frac{1}{2}\Delta_g(\ell n\, p) + \boldsymbol{h}(\ell n\, p) \\ &= \frac{1}{2}\left\{\frac{1}{p}\,\Delta_g(p) - |\operatorname{grad}_g \ell n\, p)|_g^2\right\} + \frac{1}{p}\,\boldsymbol{h}(p) \\ &= \frac{1}{p}\,\mathbb{D}(p) - \frac{1}{2}\,|\operatorname{grad}_p(\ell n\, p)|_g^2. \end{aligned}$$

Thus, (5.57) reduces to

$$\frac{dS_t}{dt} = -\int_{\mathbb{R}^n} \mathbb{D}(p_t)dx + \frac{1}{2}\,\langle |\operatorname{grad}_g \ell n\, p_t|_g^2\,\rangle_t. \tag{5.58}$$

It remains to compute the first integral in (5.58). To this end, take $v = \phi^{-1}$, $u = p_t = \phi q_t$ in (5.54) to obtain

$$p_t\mathbb{D}^*(1) - \mathbb{D}(p_t) = -\operatorname{div}[\phi q_t\{\boldsymbol{h} + \operatorname{grad}_g(\ell n\, \phi)\}].$$

Our assumptions that q_t is p.d. and $|\boldsymbol{h}|$ is a.d. imply that $|q_t \boldsymbol{h}|$ is a.d. Also, (5.43)(iii) and Lemma 5.8(iii) and (ii) imply that $|q_t \operatorname{grad}_g(\ell n\ \phi)|$ is a.d. Hence, we may apply (5.55) to conclude that

$$\int_{\mathbb{R}^n} \mathbb{D}(p_t)dx = \langle \mathbb{D}^*(1)\rangle_t. \tag{5.59}$$

But, from Proposition 5.7, $\mathbb{D}^*(1) = \frac{1}{2}\,\mathcal{L}_g(\phi) + \boldsymbol{h}(\ell n\ \phi) - \operatorname{div}_g \boldsymbol{h} = \frac{1}{2}\,\mathcal{L}_g(\phi) - \operatorname{div} \boldsymbol{h}$ (by (5.32)). Thus, combining (5.58) and (5.59), we have proved

Theorem 5.9. *Let the metric* g_{ij} *satisfy the assumptions* (5.43), *and let* $\mathbb{D} = \frac{1}{2}\,\Delta_g + \boldsymbol{h}$ *for some* C^1 *vector field* $\boldsymbol{h}$ *with* $|\boldsymbol{h}|$ *and* div $\boldsymbol{h}$ a.d. *Suppose that* $p(x,t) = \phi(x)q(x,t)$ *is a strictly positive solution of the forward diffusion equation* $(\mathbb{D}^* - (\partial/\partial t))p = 0$, *and suppose that* $q(x,t)$, $\ell n\ q(x,t)$ *and their first and second order (space) partial derivatives are* p.d. *Then the rate of change on entropy* (5.50), *is given by*

$$\frac{dS_t}{dt} = \frac{1}{2}\,\langle |\operatorname{grad}_g(\ell n\, p_t)|_g^2\rangle_g - \langle \mathcal{M}_g(\boldsymbol{h})\rangle_t,$$

where $\mathcal{M}_g(\boldsymbol{h}) = \frac{1}{2}\,\mathcal{L}_g(\phi) - \operatorname{div} \boldsymbol{h}$, *and* $\mathcal{L}_g(\phi)$ *is given by* (5.37).

It is now straightforward to interpret Theorem 5.9. The expression $\mathcal{M}_g(\boldsymbol{h})$ splits into two parts, $\frac{1}{2}\,\mathcal{L}_g(\phi)$ which is an *invariant of the diffusion geometry only,* and div h, which is an *invariant of the drift only.* (In particular, if $\boldsymbol{h}$ preserves *Euclidean* volumes, then $\mathcal{M}_g(\boldsymbol{h}) = \frac{1}{2}\mathcal{L}_g(\phi)$ is independent of $\boldsymbol{h}$.) From the formula in Theorem 5.9 , we conclude that if $\mathcal{M}_g(\boldsymbol{h}) \leq 0$ *everywhere, dS_t/dt is always positive so that entropy always increases with time.* Further, *the more negative, $\mathcal{M}_g(\boldsymbol{h})$, the faster S_t increases.* However, if $\mathcal{M}_g(\boldsymbol{h}) > 0$ *everywhere, then it is possible for entropy to decrease with time* (at the very least, it increases less rapidly than it would if the geometry and drift had no influence other than through the density p_t itself). Further, *the more positive $\mathcal{M}_g(\boldsymbol{h})$ is, the closer the system is to one for which the entropy will decrease over some time interval.* Thus, in the positive case, we see that it is possible for the degree of uncertainty about the state of the system at time t (given some initial density p_0) to *decrease* with t, so that the system progresses to a state of *greater order or canalization,* all this, of course, is in the absence of external forces ($V = 0$). ∎

Now recall Example 4.1 or (4.152) that a metric g_{ij} on $\mathbb{R}^n$ is *conformally flat* if there is a smooth function ψ such that

$$g_{ij}(x) = e^{2\psi(x)}\delta_{ij} \tag{5.60}$$

in standard co-ordinates. We shall compute the invariant $\mathcal{L}_g(\phi)$ of (5.37) for such a metric.

We have

$$\phi(x) = \sqrt{g}\,(x) = e^{n\psi(x)} \tag{5.61}$$

and the Christoffel symbols are easily computed to be

$$\begin{aligned} \Gamma^i_{ii} &= \partial_i\psi = -\Gamma^i_{jj} && \text{if} \quad i \neq j \\ \Gamma^i_{ij} &= \Gamma^i_{ji} = \partial_j\psi && \text{if} \quad i \neq j \\ \Gamma^i_{jk} &= 0 && \text{otherwise.} \end{aligned} \tag{5.62}$$

We shall compute $\mathcal{L}_g(\phi)$ in terms of the*Riemannian scalar curvature, $\mathcal{R}$.* To this end, note that

$$\mathcal{R} = g^{ij}\mathcal{R}_{ij}, \tag{5.63}$$

where $\mathcal{R}_{ij}$ is the *Ricci tensor* (4.150), is given by

$$\mathcal{R}_{ij} = \partial_k\Gamma^k_{ij} - \Gamma^r_{ik}\Gamma^k_{rj} + \Gamma^k_{ij}\partial_k(\ell\mathrm{n}\,\sqrt{g}) - \partial_i\partial_j(\ell\mathrm{n}\,\sqrt{g}). \tag{5.64}$$

Use of (5.61), (5.62), (5.63), and (5.64) now yield

$$\mathcal{R} = -(n-1)e^{-2\psi}\sum_{i=1}^{n}\{2(\partial_i\partial_i\psi) + (n-2)(\partial_i\psi)^2\}. \tag{5.65}$$

Using (5.61), and (5.62) we find that

$$\Delta_g(\ell n\ \phi) = ne^{-2\psi}\sum_{i=1}^{n}\{\partial_i\partial_i\psi + (n-2)\partial_i\psi)^2\}$$

and

$$|\text{grad}_g \ell n\ \phi|_g^2 = n^2 e^{-2\psi}\sum_{i=1}^{n}(\partial_i\psi)^2. \tag{5.66}$$

Combining (5.65), (5.66) and using (5.37), we find an invariant closely related to *quadratic dispersion* (5.8)

$$\mathcal{L}_g(\phi) = \frac{n}{2}\left\{(n+2)\ |\text{grad}_g\psi|_g^2 + \frac{\mathcal{R}}{n-1}\right\}. \tag{5.67}$$

Thus, we see that, in view of Theorem 5.9 and the remarks following, the rate of production of entropy for a conformally flat metric, and hence the possibility for the production of order, is strongly influenced by *the sign* of Riemannian scalar curvature $\mathcal{R}$. Thus (on the surfaces $|\text{grad}_g\psi|_g$ = constant), $\mathcal{L}_g(\phi)$ is *smaller* in regions where $\mathcal{R}$ is *negative, and larger* in regions where $\mathcal{R}$ is *positive.* To obtain a negative rate of change of entropy, we require $\mathcal{L}_g(\phi)$ to be large and positive. Thus, *for fixed drift vector* $\mathbf{h}$*, the more positive* $\mathcal{R}$ *is, the smaller the rate of entropy production, whereas, the more negative* $\mathcal{R}$ *is, the greater the entropy production per unit time.*

It is interesting to compare the above remarks with results on filtering (4.158), where the formula (5.67) occurs explicitly. There it is shown that, for a given signal process, the *estimation problem becomes more difficult as* $\mathcal{R}$ *becomes more negative.* In particular, in relation to the chemical ecology of communities of sessile organisms, *the negative of the quadratic dispersion,* $\mathcal{V}$, is a measure of *community vigour*, so that the more vigorous communities are harder to estimate (with a given observations process). *Regarding entropy as a measure of the degree of uncertainty, we see that the amount of uncertainty inherent in a community increases faster for more vigorous communities*, and so, from this point of view, it is not surprising that it becomes more difficult to estimate (i.e., less information is obtained from a given observation). Thus, the approach in Section 5.4 and that given here, reinforce each other in this context and show the profound role of *quadratic dispersion* in these problems.

To further justify the above discussion, we must consider when our metrics (5.60) satisfy the assumptions of Theorem 5.9, i.e., satisfy the conditions (5.43).

Observe first that the conditions on ϕ and $\ell n\ \phi$ (5.43(iii) and (iv)) imply that ψ must satisfy

$$-C|x|^k \leq \psi(x) \leq -K|x|^\sigma \quad \text{as} \quad |x| \uparrow \infty, \tag{5.68}$$

for some positive constants C, K, σ and non-negative integer k. In particular, we must have $g_{ij}(x) \to 0$ as $|x| \uparrow \infty$, and so the $g_{ij}(x)$ are *bounded*. Hence, (5.68) implies that (5.43(i)) is satisfied.

Next, observe that $\partial_k(g^{ij}) = -2(\partial_k\psi)g^{ij}$. Thus, to conform with (5.43(iii)) we must require that $\partial_k\psi$ be p.d., and hence by Lemma 5.8(ii), $\partial_k(g^{ij})$ is a.d. if and only if g^{ij} is a.d. But (5.60) and (5.61) imply that

$$g^{ij}(x) = \phi^{-2/n}(x)\delta^{ij}, \tag{5.69}$$

and so g^{ij} *is* a.d. *provided* $n > 2$. Thus, all hypotheses (5.43) are satisfied provided (5.68) holds and each $\partial_k\psi$, $\partial_k\partial_\ell\psi$ is p.d., if $n > 2$.

For $n = 2$, we cannot use Theorem 5.9 as it stands. The essential point about a.d. functions used is that such functions belong to $L^1(\mathbb{R}^n; d\mu)$. To retain this property, and obtain a formula for dS_t/dt like that in Theorem 5.9, we must impose very strict domination conditions on the *solutions* $q(x,t)$ of the diffusion equation $(\mathbb{D}^+ - \partial/\partial t)q = 0$, to compensate for the lack of a good constraint on the g^{ij}. Thus, we can recover the results of Theorem 5.9, *if we assume that* $\ell n\ q_t$ *is* p.d. (as before) *and that* q_t *and its first and second order partial derivatives are dominated by*

$$C(1 + |x|^k)\phi^\varepsilon(x) \tag{5.70}$$

for some $\varepsilon > 0$. The proof is essentially the same. The point is that the product of g^{ij} with a function dominated by (5.70) is a.d.

Note that solutions $q(x,t)$ dominated by (5.70) tend to zero as $|x| \uparrow \infty$, so *that the result for* $n = 2$ *applies to a very much more restricted class of solutions than the result for* $n > 2$. ▮

Recall that, from Proposition 5.6, ϕ is a *stationary density* for the diffusion operator $\mathbb{D}^*$ if $V = \operatorname{div}_g \boldsymbol{h}$; i.e., if $\mathbb{D} = \frac{1}{2}\Delta_g + \boldsymbol{h} + \operatorname{div}_g \boldsymbol{h}$. For a conformally flat metric (5.60), this density is given by (5.61). We shall be particularly interested in the case in which (5.60) is a *Gaussian metric*; i.e., a metric for which

$$\psi(x) = -\frac{1}{2}(x-\mu)A(x-\mu)^* + c \tag{5.71}$$

for some $\mu \in \mathbb{R}^n$, constant c, and positive definite, symmetric matrix $A = (a_{ij})$ (here, * denotes transpose). Such metrics are of interest in connection with the chemical ecology.

For a Gaussian metric, the normalized stationary probability density $p^{st} = \phi/V_g(\mathbb{R}^n)$, is the given by the n-variate Gaussian distribution with mean μ and covariance matrix nA,

$$p^{st}(x) = \left(\frac{n}{2\pi}\right)^{n/2}(\det A)^{1/2}\exp\left\{-\frac{n}{2}(x-\mu)A(x-\mu)^*\right\}. \tag{5.72}$$

We shall consider the *entropy* of (5.72) in terms of the "temperature-like" parameter

$$T = (\det A)^{-1/n}, \tag{5.73}$$

i.e., *T is the reciprocal of the geometric mean of the eigenvalues of* A. By analogy with classical thermodynamics, we shall refer to T as the *absolute temperature* of the Gaussian metric defined by (5.71). We shall show that, for fixed n, *the entropy is a function of T alone.*

Performing an orientation preserving orthogonal transformation of coordinates, if necessary, we may suppose that $A = \operatorname{diag}(\lambda_1, \ldots, \lambda_n)$, and then, via (5.73), (5.72) reduces to the product

$$p^{st}(x) = \left(\frac{n}{2\pi T}\right)^{n/2}\prod_{i=1}^{n}\exp\left\{\frac{n\lambda_i}{2}(x^i-\mu^i)^2\right\}, \tag{5.74}$$

so, the entropy S is given by

$$\begin{aligned} S &= -\int_{\mathbb{R}^n} p^{st}(x)\ell n\left(p^{st}(x)\right)dx \\ &= -\frac{n}{2}\ell n\left(\frac{2\pi T}{n}\right) - \left(\frac{n}{2\pi T}\right)^{n/2}\sum_{i=1}^{n}\int_{\mathbb{R}^n}\frac{n\lambda_i}{2}(x^i-\mu^i)^2 \\ &\quad \prod_{j=1}^{n}\exp\left\{-\frac{n\lambda_j}{2}(x^j-\mu^j)^2\right\}dx. \end{aligned}$$

The integral is easily computed to be

$$\frac{1}{2}\sqrt{2\pi/n\lambda_i}\cdot\prod_{j\neq i}\sqrt{2\pi/n\lambda_j}=\frac{1}{2}\,(2\pi T/n)^{n/2},$$

using (5.73). Thus, we obtain an expression for S as a function of T,

$$S=\frac{n}{2}\left\{1+\ell n\left(\frac{2\pi T}{n}\right)\right\}. \tag{5.75}$$

Note that S is an increasing function of T. Thus, large temperatures means more disorder, while small temperatures mean more order.

Also of interest is the variation of S with n for fixed T. Replacing n by a continuous positive variable u, we find that $S(u)$ has a maximum value of T at $u=2\pi T$, and $S(u)$ increases as u increases from 0 to $2\pi T$, and decreases as u increases beyond $2\pi T$. In fact, $S(u)=0$ when $u=2\pi Te$, and becomes negative as u increases beyond this point. It follows that, *if $n\geq 2\pi T$, the more species there are living in a community at a given temperature, the more orderly the system is.* Furthermore, this effect is enhanced if the temperature *decreases* as more species are added, and is diluted otherwise. This type of result, with identical parities, was proved for the *vigour* $\mathcal{V}$ in the filtering sections.

Remark 5.2. We note finally, that using the formulas (5.66), we may compute the invariant $\mathcal{L}_g(\phi)$ for the Gaussian metric (5.71). We find

$$\mathcal{L}_g(\phi)=ne^{-2\psi(x)}\{2|(x-\mu)A|^2+\operatorname{Tr}(A)\},$$

where $\operatorname{Tr}(A)=$ trace of A. It is clear then, that $\mathcal{L}_g(\phi)$ depends on *all* the eigenvalues of A separately, and is not a simple function of T alone. Note also that

$$\mathcal{L}_g(\phi)\geq ne^{-2\psi(x)}\operatorname{Tr}(A)>0$$

for all x, since A is positive definite. Thus, from the discussion after the proof of Theorem 5.9, we see that *the rate of entropy production is strictly smaller for a Gaussian metric than for a metric with non-positive $\mathcal{L}_g(\phi)$ but comparable density at some given time t.*

Chapter 6

Diffusion on the Tangent and Indicatrix Bundles

Finslerian diffusion has been constructed in Chapter 3 by extending the notion of stochastic development from Riemannian manifolds to Finsler manifolds. In the present chapter we shall study Finslerian diffusion as a process on the slit tangent bundle $\mathring{T}M$ and the indicatrix bundle IM of a Finsler manifold M.

In Section 6.1 it is demonstrated that the Finslerian hv-Brownian motion introduced in Chapter 3 is equivalent to Riemannian Brownian motion with drift on the slit tangent bundle $\mathring{T}M$ equipped with a suitable Riemannian metric, the so-called Sasaki lift or diagonal lift of the Finsler metric. In Section 6.3 an alternative approach to Finslerian diffusion will be developed with $\mathring{T}M$ replaced by the indicatrix bundle IM, in line with the view that Finslerian objects depend in fact only on the direction of a tangent vector to M because of their homogeneity in $y \in \mathring{T}M$. This leads to the definition of an indicatrized Finsler hv-Brownian motion. Similarly as in the case of hv-Brownian motion on $\mathring{T}M$, in Section 6.4 it is demonstrated that the indicatrized hv-Brownian motion is equivalent to a Riemannian Brownian motion with drift on the indicatrix bundle IM equipped with the diagonal lift metric. The drift on $\mathring{T}M$ and the drift on IM can both be expressed in terms of the torsion tensors P and C of the Cartan connection.

Throughout this chapter we shall assume that M is an n-dimensional Finsler manifold with metric function L, positive definite metric tensor g_{ij} and the Cartan connection $(\mathbf{N}^i_j, \mathbf{F}^i_{jk}, \mathbf{C}^i_{jk})$. Given a local coordinate system (x^i) on M, we take the induced coordinates (x^i, y^i) on $\mathring{T}M$ and consider the adapted frame $\delta_i, \partial_{(i)}$ on $\mathring{T}M$, where

$$\delta_i = \partial_i - \mathbf{N}_i^j(x,y)\partial_{(j)}.$$

Here $\partial_i = \partial/\partial x^i$ and $\partial_{(i)} = \partial/\partial y^i$. For ease of notation, in the present chapter we shall write $\partial_{(i)}$ in place of the symbol $\dot{\partial}_i$ used throughout the rest of this book. The dual frame to $\delta_i, \partial_{(i)}$ will be denoted by $dx^i, \delta y^i$, where

$$\delta y^i = dy^i + \mathbf{N}_j^i(x,y)dx^j.$$

6.1 Slit Tangent Bundle as Riemannian Manifold

The goal of the present section is to demonstrate that Finslerian *hv*-Brownian motion as defined in Chapter 3 by means of Finslerian stochastic development can also be regarded as Riemannian Brownian motion with a suitable drift on the slit tangent bundle $\mathring{T}M$ equipped with the so-called diagonal lift metric, which turns $\mathring{T}M$ into a Riemannian manifold. The drift can be expressed in terms of the torsion tensors P^i_{jk} and C^i_{jk} of the Cartan connection.

Definition 6.1. The *diagonal lift metric* on $\mathring{T}M$ is defined by

$$G(x,y) = g_{ij}(x,y)dx^i \otimes dx^j + g_{ij}(x,y)\delta y^i \otimes \delta y^j,$$

where g is the Finsler metric tensor.

If g is positive definite, then so is G. Equipped with the metric tensor G, the slit tangent bundle $\mathring{T}M$ becomes a Riemannian manifold. We shall also need the following connection on $\mathring{T}M$.

Definition 6.2. The connection ∇ on $\mathring{T}M$ such that

$$\begin{aligned}
\nabla_{\delta_i}\delta_j &= \Gamma^k_{ij}\delta_k + \Gamma^{(k)}_{ij}\partial_{(k)} = \mathbf{F}^k_{ij}\delta_k,\\
\nabla_{\delta_i}\partial_{(j)} &= \Gamma^k_{i(j)}\delta_k + \Gamma^{(k)}_{i(j)}\partial_{(k)} = \mathbf{F}^k_{ij}\partial_{(k)},\\
\nabla_{\partial_{(i)}}\delta_j &= \Gamma^k_{(i)j}\delta_k + \Gamma^{(k)}_{(i)j}\partial_{(k)} = \mathbf{C}^k_{ij}\delta_k,\\
\nabla_{\partial_{(i)}}\partial_{(j)} &= \Gamma^k_{(i)(j)}\delta_k + \Gamma^{(k)}_{(i)(j)}\partial_{(k)} = \mathbf{C}^k_{ij}\partial_{(k)}
\end{aligned}$$

is called the *horizontal lift of the Cartan connection to* $\mathring{T}M$.

The above formulae define the coefficients $\Gamma^k_{ij}, \Gamma^{(k)}_{ij}, \Gamma^k_{(i),j}$, etc. of ∇ in the adapted frame $\delta_k, \partial_{(k)}$ on $\mathring{T}M$ in terms of $\mathbf{F}^k_{ij}$ and $\mathbf{C}^k_{ij}$. Namely,

$$\Gamma^k_{ij} = \Gamma^{(k)}_{i(j)} = \mathbf{F}^k_{ij},$$
$$\Gamma^k_{(i)j} = \Gamma^{(k)}_{(i)(j)} = \mathbf{C}^k_{ij},$$
$$\Gamma^{(k)}_{ij} = \Gamma^k_{i(j)} = \Gamma^{(k)}_{(i)j} = \Gamma^k_{(i)(j)} = 0.$$

Proposition 6.1. *∇ is a metrical connection with respect to the Riemannian metric G on $\mathring{T}M$, that is,*

$$\nabla G = 0.$$

Proof. The proposition follows immediately from the definitions of ∇ and G, and the fact that the Cartan connection is h- and v−metrical. ∎

Even though ∇ is a metrical connection, it differs from the Levi-Civita connection in that it has non-vanishing torsion, which is computed in the next proposition.

Proposition 6.2. *In the adapted frame $\delta_i, \partial_{(i)}$ the torsion tensor T of ∇ can be expressed as follows:*

$$\begin{aligned}
T(\delta_i, \delta_j) &= T^k_{ij}\delta_k + T^{(k)}_{ij}\partial_{(k)} = R^i_{jk}\partial_{(k)},\\
T(\delta_i, \partial_{(j)}) &= T^k_{i(j)}\delta_k + T^{(k)}_{i(j)}\partial_{(k)} = -C^k_{ij}\delta_k - P^k_{ij}\partial_{(k)},\\
T(\partial_{(i)}, \delta_j) &= T^k_{(i)j}\delta_k + T^{(k)}_{(i)j}\partial_{(k)} = +C^k_{ij}\delta_k + P^k_{ij}\partial_{(k)},\\
T(\partial_{(i)}, \partial_{(j)}) &= T^k_{(i)(j)}\delta_k + T^{(k)}_{(i)(j)}\partial_{(k)} = 0.
\end{aligned}$$

where

$$\begin{aligned}
R^i_{jk} &= \delta_j \mathbf{N}^i_k - \delta_k \mathbf{N}^i_j,\\
P^i_{jk} &= \partial_{(j)}\mathbf{N}^i_k - \mathbf{F}^i_{kj}, \qquad (6.1)\\
C^i_{jk} &= \mathbf{C}^i_{jk}
\end{aligned}$$

are the coefficients of the three non-vanishing torsion tensors R, P, C of the Cartan connection in the adapted frame $\delta_i, \partial_{(i)}$.

Proof. The formulae follow by direct computation from the formulae for the coefficients ∇ in Definition 6.2 and the general definition

$$T(X,Y) = \nabla_X Y - \nabla_Y X - [X,Y]$$

of the torsion tensor. ∎

6.2 hv-Development as Riemannian Development with Drift

According to Propositions 6.1 and 6.2 the horizontal lift ∇ of the Cartan connection is a metrical connection with torsion on $\mathring{T}M$ regarded as a Riemannian manifold with the diagonal lift metric G. The following theorem extends the results of Section 3.1.2 to the case when the Levi-Civita connection γ^i_{jk} on a Riemannian manifold is replaced by a metrical connection Γ^i_{jk} with torsion.

Theorem 6.3. *Let M be an n-dimensional Riemannian manifold with metric tensor g_{ij} and let Γ^i_{jk} be the coefficients of a metrical connection with torsion tensor*

$$T^i_{jk} = \Gamma^i_{jk} - \Gamma^i_{kj}.$$

Then for any diffusion (X, Z) on the orthonormal frame bundle OM satisfying the system of SDEs

$$\begin{aligned} dX^i &= Z^i_m \circ dW^m, \\ dZ^i_m &= -\Gamma^i_{jk}(X) Z^j_m Z^k_l \circ dW^l, \end{aligned} \tag{6.2}$$

X is a diffusion on M with generator

$$D = \frac{1}{2} g^{ij}(\partial_i \partial_j - \Gamma^k_{ij} \partial_k) = \frac{1}{2}\Delta + B, \tag{6.3}$$

where

$$\Delta = g^{ij}(\partial_i \partial_j - \gamma^k_{ij} \partial_k)$$

is the Laplace-Beltrami operator and B a vector field on M such that

$$B = \frac{1}{2} g^{ik} T^j_{jk} \partial_i. \tag{6.4}$$

This is to say, X is a Riemannian Brownian motion on M with drift B given by (6.4). Note that Frame fields (Z^i_m) are denoted as upper case Roman letters in this chapter.

Proof. This result is well known. For the proof we refer, for example, to [IW89]. ∎

We are now in a position to present the main result of this section, according to which Finslerian hv-Brownian motion can be regarded as Riemannian Brownian motion with drift on $\mathring{T}M$.

Theorem 6.4. *A diffusion (X, Y) on $\mathring{T}M$ is a Finslerian hv-Brownian motion if and only if it is a Riemannian Brownian motion with drift*

$$B = \frac{1}{2} g^{ik} P^j_{jk} \delta_i - \frac{1}{2} g^{ik} C^j_{kj} \partial_{(i)} \tag{6.5}$$

on the slit tangent bundle $\mathring{T}M$ regarded as a Riemannian manifold with the horizontal lift metric G of Definition 6.1. Here P^i_{jk} and C^i_{jk} are the coefficients (6.1) *of the torsion tensors P and C of the Cartan connection in the adapted frame $\delta_i, \partial_{(i)}$.*

Proof. It suffices to demonstrate that the generator D of Riemannian Brownian motion on $\mathring{T}M$ with drift (6.5) is the same as that of the Finslerian hv-Brownian motion. To this end we apply Theorem 6.3 to the case of $\mathring{T}M$ with the diagonal lift metric G and the horizontal lift ∇ of the Cartan connection to $\mathring{T}M$. The drift B in Theorem 6.3 can be written as

$$\begin{aligned} B &= \frac{1}{2} g^{ik} (T^j_{jk} + T^{(j)}_{(j)k}) \delta_i + \frac{1}{2} g^{ik} (T^j_{j(k)} + T^{(j)}_{(j)(k)}) \partial_{(i)} \\ &= \frac{1}{2} g^{ik} P^j_{jk} \delta_i - \frac{1}{2} g^{ik} C^j_{kj} \partial_{(i)}. \end{aligned}$$

This follows immediately from Proposition 6.2. Now, expressing the generator

$$D = \frac{1}{2} \Delta + B$$

in Theorem 6.3 in terms of the adapted frame $\delta_i, \delta_{(i)}$ and applying the formulae for the connection coefficients of ∇ given in Definition 6.2, we find that

$$D = \frac{1}{2} g^{ij} (\delta_i \delta_j - \mathbf{F}^k_{ij} \delta_k) + \frac{1}{2} g^{ij} (\partial_{(i)} \partial_{(j)} - \mathbf{C}^k_{ij} \partial_{(k)}).$$

This is the generator of Finslerian hv-Brownian motion, see Theorem 3.11. ∎

6.3 Indicatrized Finslerian Stochastic Development

In the present section we adopt the point of view that Finslerian objects should depend only on the direction of a tangent vector $y \in \mathring{T}M$, the dependence on the length of y being completely determined by the homogeneity conditions. In line with this point of view we reformulate the theory of Finslerian diffusion, using the indicatrix bundle IM with fibres

$$IM_x = \{y \in TM_x : L(x,y) = 1\}$$

in place of the slit tangent bundle $\mathring{T}M$.

While the system (3.39) of SDEs defines Finslerian hv-stochastic development in terms of 'rolling' along a Brownian motion (W, V) on the 'tangent bundle' $\mathbf{R}^n \times \mathbf{R}^n$ over $\mathbf{R}^n$, we shall introduce 'rolling' along Brownian motion (W, U) on the 'sphere bundle' $\mathbf{R}^n \times S^{n-1}$ over $\mathbf{R}^n$, where

$$S^{n-1} = \{x \in \mathbf{R}^n : \delta_{ij}x^i x^j = 1\}$$

is the unit sphere in $\mathbf{R}^n$. Thus, we now assume that W is a standard Brownian motion on $\mathbf{R}^n$ and U is a Riemannian Brownian motion on the unit sphere $S^{n-1} \subset \mathbf{R}^n$ such that U and W are independent.

We consider the unit sphere $S^{n-1} \subset \mathbf{R}^n$ as a Riemannian manifold with metric induced from $\mathbf{R}^n$. By definition, a Riemannian Brownian motion U on S^{n-1} is an S^{n-1}-valued diffusion with generator $D = \frac{1}{2}\Delta$, where Δ is the Laplace-Beltrami operator on S^{n-1}, which, as is well known, can be written as

$$\Delta = (\delta^{ij} - x^i x^j)\partial_i \partial_j - (n-1)x^k \partial_k,$$

where x^i are the canonical Cartesian coordinates on $\mathbf{R}^n$ and $\partial_i = \partial/\partial x^i$. Even though expressed in terms of the operators ∂_i on $\mathbf{R}^n$, the above expression defines an operator on S^{n-1}, since, for any smooth functions $f_1, f_2 : \mathbf{R}^n \to \mathbf{R}$, if $f_1 = f_2$ on S^{n-1}, then $\Delta f_1 = \Delta f_2$ on S^{n-1}.

To construct a concrete specimen of Riemannian Brownian motion U on S^{n-1}, one can employ the system (3.17) of SDEs for Riemannian stochastic development. However, for our purposes, it proves more convenient to use another representation of Brownian motion on S^{n-1} due to Stroock [Str71] (see also [IW89], Section III.2). Namely, let

$$A^i_j(x) = \delta^i_j - \delta_{jk}x^i x^k/|x|^2$$

for any $0 \neq x \in \mathbf{R}^n$ and let V be a standard Brownian motion on $\mathbf{R}^n$. Then the solution U to the SDE

$$dU^i = A^i_j(U) \circ dV^j \tag{6.6}$$

with initial condition $U(0) \in S^{n-1}$ is a Brownian motion on S^{n-1}. It is easily verified that

$$\begin{aligned} dU^i &= (\delta^i_j - \delta_{ij}U^iU^k)dV^j - \frac{1}{2}(n-1)U^i dt, \\ dU^i dU^j &= (\delta^{ij} - U^iU^j)dt. \end{aligned} \tag{6.7}$$

Definition 6.3. Let (X, Y, Z) be a solution to the system

$$\begin{aligned} dX^i &= Z^i_j \circ dW^j \\ \delta Y^i &= Z^i_j \circ dU^j \\ dZ^i_j &= -\mathbf{F}^i_{kl}(X,Y)Z^l_j \circ dX^k - \mathbf{C}^i_{kl}(X,Y)Z^l_j \circ \delta Y^k \end{aligned} \tag{6.8}$$

of SDEs, where

$$\delta Y^i = dY^i + \mathbf{N}^i_j(X,Y) \circ dX^j$$

and where W is a standard Brownian motion on $\mathbf{R}^n$ and U a Riemannian Brownian motion on the unit sphere $S^{n-1} \subset \mathbf{R}^n$ such that U and W are independent, with initial conditions $X(0) = x_0 \in M$, $Y(0) = y_0 \in IM_{x_0}$, and $Z(0) = z_0$, a frame in TM_{x_0} orthonormal with respect to the quadratic form $g(x_0, y_0)$ such that $Y^i(0) = U^j(0)Z^i_j(0)$. Then (X, Y, Z) is called the *indicatrized Finslerian hv-development.*

As compared to Definition 3.2, we have used a Riemannian motion U on the unit sphere S^{n-1} in place of a Euclidean Brownian motion in $\mathbf{R}^n$, and we introduced the initial condition $Y^i(0) = U^j(0)Z^i_j(0)$. As the theorem below demonstrates, the resulting process (X, Y, Z) is restricted to a subbundle of the orthonormal Finsler bundle OTM.

Definition 6.4. The orthonormal Finsler bundle $OIM \subset OTM$ over $IM \subset \mathring{T}M$ is, by definition, a subbundle of OTM induced by the projection $\pi : OTM \to \mathring{T}M$, that is, $OIM = \pi^{-1}(IM)$.

Theorem 6.5. *The indicatrized Finslerian hv-development (X, Y, Z) is a diffusion on the orthonormal Finsler bundle OIM over IM, and (X, Y) is a diffusion on IM with generator*

$$D = \frac{1}{2}g^{ij}(\delta_i\delta_j - \mathbf{F}^k_{ij}\delta_k) + \frac{1}{2}(g^{ij} - y^iy^j)(\partial_{(i)}\partial_{(j)} - \mathbf{C}^k_{ij}\partial_{(k)}) - \frac{1}{2}(n-1)y^k\partial_{(k)}. \tag{6.9}$$

Remark 6.1. Even though the generator D given by (6.9) is expressed in terms of the adapted frame $\delta_i, \partial_{(i)}$ on $\mathring{T}M$, it can be regarded as an operator over IM in the sense that if $f_1, f_2 : \mathring{T}M \to \mathbf{R}$ are smooth functions such that $f_1 = f_2$ on IM, then $Df_1 = Df_2$ on IM. This property of D can be verified by direct computation, but it also follows from the proof of Theorem 6.5 as well as from assertion b) of Theorem 6.9

Lemma 6.6. *Under the assumptions of Theorem 6.5,*

a) $Y^i = U^j Z^i_j \quad P$-a.s.,

b) $g_{ij}(X,Y)Z^i_k Z^j_l = \delta_{kl} \quad P$-a.s.,

c) $L(X,Y) = 1 \quad P$-a.s..

Proof. a) Given the solution (X,Y,Z) to (6.8) with the initial conditions described in Theorem 6.5, we consider the SDE

$$dA^i = -\mathbf{F}^i_{kl}(X,Y)A^l \circ dX^k - \mathbf{C}^i_{kl}(X,Y)A^l \circ \delta Y^k + Z^i_j \circ dU^j. \tag{6.10}$$

Since $\mathbf{C}^i_{jk}(x,y)y^k = 0$ and the Cartan connection is deflection-free, i.e., $\mathbf{F}^i_{jk}(x,y)y^k = \mathbf{N}^i_j(x,y)$, it follows from the second equation in (6.7) that $A^i = Y^i$ is a solution to (6.10). By the Itô formula, $d(U^j Z^i_j) = U^j \circ dZ^i_j + Z^i_j \circ dU^j$. Substituting for dZ^i_j from the third equation in (6.8), we find that $A^i = U^j Z^i_j$ is also a solution to (6.10). But, by assumption, $Y^i(0) = U^j(0)Z^i_j(0)$, which, by uniqueness, means that $Y^i = U^j Z^i_j$, as required.

b) The proof of assertion b) is exactly the same as that of (Prop. 3.9) above.

c) By the Itô formula,

$$dL^2(X,Y) = \delta_i L^2(X,Y) \circ dX^i + \partial_{(i)} L^2(X,Y) \circ \delta Y^i.$$

Since the Cartan connection is deflection-free, it follows that $\delta_i L^2(x,y) = 0$. We also have $\partial_{(i)} L^2(x,y) = 2g_{ij}(x,y)y^j$. Thus, by the second equation in (6.7) and assertions a) and b) of the present lemma,

$$\begin{aligned} dL^2(X,Y) &= 2g_{ij}(X,Y)Y^i \circ \delta Y^j \\ &= 2g_{ij}(X,Y)U^k Z^i_k Z^j_l \circ dU^l \\ &= 2\delta_{kl} U^k \circ dU^l \\ &= 2d(\delta_{kl} U^k U^l) = 0, \end{aligned}$$

for U is a Brownian motion on the unit sphere, i.e., $\delta_{kl} U^k U^l = 1$. Because, by assumption, $L(X(0), Y(0)) = 1$, it follows that $L(X,Y) = 1$, as required. ∎

Proof of Theorem 6.5. As a consequence of assertions b) and c) of Lemma 6.6, (X, Y, Z) is a diffusion on the orthonormal Finsler bundle OIM over IM with projection (X, Y) onto IM.

Let us now compute the stochastic differentials dX^i and δY^i from (6.7):

$$\begin{aligned} dX^i &= Z^i_j dW^j - \frac{1}{2}\mathbf{F}^i_{kl}(X,Y)Z^k_m Z^l_j dW^m dW^j \\ &\quad -\frac{1}{2}\mathbf{C}^i_{kl}(X,Y)Z^k_m Z^l_j dU^m dW^j, \\ \delta Y^i &= Z^i_j dU^j - \frac{1}{2}\mathbf{F}^i_{kl}(X,Y)Z^k_m Z^l_j dW^m dU^j \\ &\quad -\frac{1}{2}\mathbf{C}^i_{kl}(X,Y)Z^k_m Z^l_j dU^m dU^j. \end{aligned}$$

But $dW^m dW^j = \delta^{mj}dt$ by the Itô rules, and $dU^m dW^j = dW^m dU^j = 0$ because U and W are independent. Without loss of generality, we can assume that U is the representation of Brownian motion on the unit sphere $S^{n-1} \subset \mathbf{R}^n$ defined by (6.6), which means that we can use formulae (6.7) for dU^j and $dU^m dU^j$. It follows that

$$\begin{aligned} dX^i &= Z^i_j dW^j - \frac{1}{2}\mathbf{F}^i_{kl}(X,Y)Z^k_m Z^l_j \delta^{mj} dt \\ &= Z^i_j dW^j - \frac{1}{2}g^{kl}(X,Y)\mathbf{F}^i_{kl}(X,Y)dt, \\ \delta Y^i &= Z^i_j(\delta^j_k - \delta_{kl}U^j U^l)dV^k - \frac{1}{2}(n-1)U^j Z^i_j dt \\ &\quad -\frac{1}{2}\mathbf{C}^i_{kl}(X,Y)Z^k_m Z^l_j(\delta^{mj} - U^m U^j)dt \\ &= Z^i_j(\delta^j_k - \delta_{kl}U^j U^l)dV^k - \frac{1}{2}(n-1)Y^i dt \\ &\quad -\frac{1}{2}g^{kl}(X,Y)\mathbf{C}^i_{kl}(X,Y)dt, \end{aligned}$$

since $Z^k_m Z^l_j \delta^{mj} = g^{kl}(X,Y)$ by assertion b) of Lemma 6.6, $Y^i = U^j Z^i_j$ by assertion a) of Lemma 6.6, and $\mathbf{C}^i_{kl}(x,y)y^l = 0$. It also follows that

$$\begin{aligned} &dX^i dX^j = Z^i_k Z^j_l dW^k dW^l = Z^i_k Z^j_l \delta^{kl} dt = g^{ij}(X,Y)dt, \\ &\delta Y^i \delta Y^j = Z^i_k Z^j_l dU^k dU^l = Z^i_k Z^j_l(\delta^{kl} - U^k U^l)dt \\ &\qquad = (g^{ij}(X,Y) - Y^i Y^j)dt, \\ &dX^i \delta Y^j = Z^i_k Z^j_l dW^k dU^l = 0, \\ &\delta Y^i dX^j = Z^i_k Z^j_l dU^k dW^l = 0, \end{aligned}$$

Thus, for any smooth function $f : \mathring{T}M \to \mathbf{R}$,

$$\begin{aligned}
df(X,Y) &= \delta_i f(X,Y) \circ dX^i + \partial_{(i)} f(X,Y) \circ \delta Y^i \\
&= \delta_i f(X,Y) dX^i + \partial_{(i)} f(X,Y) \delta Y^i \\
&\quad + \frac{1}{2}\delta_i \delta_j f(X,Y) dX^i dX^j + \frac{1}{2}\delta_i \partial_{(j)} f(X,Y) dX^i \delta Y^j \\
&\quad + \frac{1}{2}\partial_{(i)} \delta_j f(X,Y) \delta Y^i dX^j + \frac{1}{2}\partial_{(i)} \partial_{(j)} f(X,Y) \delta Y^i \delta Y^j \\
&= \delta_i f(X,Y) Z^i_j dW^j + \partial_{(i)} f(X,Y) Z^i_j (\delta^j_k - \delta_{kl} U^j U^l) dV^k \\
&\quad + [\frac{1}{2} g^{ij}(X,Y)(\delta_i \delta_j f(X,Y) - \mathbf{F}^k_{ij}(X,Y)\delta_k f(X,Y)) \\
&\quad + \frac{1}{2}(g^{ij}(X,Y) - Y^i Y^j)(\partial_{(i)}\partial_{(j)} f(X,Y) - \mathbf{C}^k_{ij}(X,Y)\partial_{(k)} f(X,Y)) \\
&\quad - \frac{1}{2}(n-1) Y^k \partial_{(k)} f(X,Y)] dt.
\end{aligned}$$

As a consequence, (X,Y) is a diffusion with generator D given by the expression in square brackets multiplying dt above. Since, by assertion c) of Lemma 6.6, $L(X,Y) = 1$, that is, $(X,Y) \in IM$, it follows from the above calculation that D is an operator on IM in the sense of Remark 6.1. ∎

Definition 6.5. The solution (X,Y,Z) to (6.8) with the initial conditions in Theorem 6.5 will be called an *indicatrized Finslerian hv-stochastic development.*

Definition 6.6. A diffusion (X,Y) on IM with generator D given by (6.9) will be called an *indicatrized Finslerian hv-Brownian motion.*

We have just proved that if (X,Y,Z) is an indicatrized *hv*-stochastic development, then (X,Y) is an indicatrized *hv*-Brownian motion. In the next section we shall demonstrate that an indicatrized *hv*-Brownian motion can also be represented as a Riemannian Brownian motion with drift on the indicatrix bundle IM equipped with the Riemannian metric induced from $\mathring{T}M$.

6.4 Indicatrized *hv*-Development Viewed as Riemannian

As in the previous section, we assume that M is an n-dimensional Finsler manifold with metric function L, positive definite metric tensor g_{ij}, and the Cartan connection $(\mathbf{N}^i_j, \mathbf{F}^i_{jk}, \mathbf{C}^i_{jk})$. We shall use the standard summation convention on upper and lower indices supplemented with the rule that there is no summation with respect to $n =$ dim M. We also assume that Latin indices i, j, k, l, m run from 1 to n, while Greek indices $\varkappa, \lambda, \mu$ run from 1 to $n-1$.

We consider the indicatrix bundle $IM \subset \overset{\circ}{T}M$ as a Riemannian manifold equipped with the metric $\widehat{G}$ induced by the diagonal lift metric G defined on $\overset{\circ}{T}M$ by Definition 6.1. We shall prove the following theorem.

Theorem 6.7. *A diffusion (X, Y) on IM is an indicatrized hv-Brownian motion if and only if it is a Riemannian Brownian motion with drift B on the indicatrix bundle IM equipped with the Riemannian metric $\widehat{G}$. The drift is expressed as*

$$B = \frac{1}{2} g^{ik} P^j_{jk} \delta_i - \frac{1}{2} g^{ik} C^j_{kj} \partial_{(i)} \tag{6.11}$$

in terms of the torsion tensors P^i_{jk} and C^i_{jk} of the Cartan connection and the adapted frame $\delta_i, \partial_{(i)}$.

Remark 6.2. The vector field B on $\overset{\circ}{T}M$ defined in Theorem 6.7 is tangent to IM, which means that it can also be regarded as a vector field on IM. This can be verified by direct computation, but it also follows from the proof of assertion a) of Theorem 6.9 below.

Theorem 6.7 will be proved at the end of the present section. Right now, let us introduce the notation and state some facts concerning the indicatrix bundle IM to be used in the remainder of this section.

Let (x^k) be local coordinates on M and let (x^k, y^k) be the induced coordinates on $\overset{\circ}{T}M$. The vector fields

$$\partial_k = \partial/\partial x^k, \quad \partial_{(k)} = \partial/\partial y^k, \quad \delta_k = \partial_k - \mathbf{N}^j_k(x, y)\partial_{(j)}$$

on $\overset{\circ}{T}M$ are defined as before. We also choose another system $(\widetilde{x}^k, \widetilde{y}^k)$ of local coordinates on $\overset{\circ}{T}M$ defined by

$$\widetilde{x}^k(x, y) = x^k, \quad \widetilde{y}^\varkappa(x, y) = y^\varkappa / L(x, y), \quad \widetilde{y}^n(x, y) = L(x, y),$$

subject to the condition $\partial_{(n)} L(x, y) \neq 0$. Since $y \neq 0$ and $\partial_{(k)} L(x, y) = y_k / L$ for any k, reshuffling the indices if necessary, one can always ensure that the latter condition is satisfied in a neighbourhood of any given point in $\overset{\circ}{T}M$. Then, by the implicit function theorem, $(\widetilde{x}^k, \widetilde{y}^k)$ are well defined local coordinates in a neighbourhood of the given point. Also, the indicatrix bundle IM is given by the equation $\widetilde{y}^n(x,y) = 1$ and $(\widetilde{x}^k, \widetilde{y}^\varkappa)$ are local coordinates on IM.

Using the identities

$$\begin{aligned} \partial_{(k)} L(x,y) &= y_k / L(x,y), \\ \partial_k L(x,y) &= \overset{0}{\gamma}_{0k}(x,y)/L(x,y), \end{aligned}$$

where $\overset{0}{\gamma}_{0k}(x,y) = \gamma^i_{jk}(x,y) y_i y^j$, one finds that

$$\begin{gathered} \frac{\partial x^k}{\partial \widetilde{x}^l} = \delta^k_l, \quad \frac{\partial x^k}{\partial \widetilde{y}^l} = 0, \quad \frac{\partial y^\varkappa}{\partial \widetilde{x}^l} = 0, \quad \frac{\partial y^n}{\partial \widetilde{x}^l} = -\frac{\overset{0}{\gamma}_{0l}}{y_n}, \\ \frac{\partial y^\varkappa}{\partial \widetilde{y}^\lambda} = \delta^\varkappa_\lambda L, \quad \frac{\partial y^n}{\partial \widetilde{y}^\lambda} = -\frac{L y_\lambda}{y_n}, \quad \frac{\partial y^k}{\partial \widetilde{y}^n} = \frac{y^k}{L}. \end{gathered} \tag{6.12}$$

We define the vector fields $\widetilde{\partial}_k = \partial/\partial\widetilde{x}^k$ and $\widetilde{\partial}_{(k)} = \partial/\partial\widetilde{y}^k$ on $\mathring{T}M$. From (6.12) it follows that

$$\widetilde{\partial}_k = \partial_k - (\overset{0}{\gamma}_{0k}/y_n)\partial_{(n)}, \quad \widetilde{\partial}_{(\varkappa)} = L\partial_{(\varkappa)} - (L y_\varkappa / y_n)\partial_{(n)}, \quad \widetilde{\partial}_{(n)} = (y^i/L)\partial_{(i)}.$$

Clearly, the vector fields $\widetilde{\partial}_k$ and $\widetilde{\partial}_{(\varkappa)}$ are tangent to IM.

Proposition 6.8.

a) The vector fields $\delta_k = \partial_k - \mathbf{N}^j_k(x,y)\partial_{(j)}$ are tangent to IM;

b) The vector field $\widetilde{\partial}_{(n)} = \partial/\partial\widetilde{y}^n$ is perpendicular to IM in the diagonal lift metric G.

Proof. The proof depends on the condition $\mathbf{N}^j_k(x,y) y_j = \overset{0}{\gamma}_{0k}(x,y)$ satisfied by the nonlinear part $\mathbf{N}^j_k$ of the Cartan connection.

To prove assertion a), it suffices to express δ_k as a linear combination of $\widetilde{\partial}_k$ and $\widetilde{\partial}_{(\varkappa)}$. We get

$$\begin{aligned} \delta_k &= \partial_k - \mathbf{N}^l_k \partial_{(l)} \\ &= \widetilde{\partial}_k - (\mathbf{N}^\lambda_k / L)\widetilde{\partial}_{(\lambda)} + [(\overset{0}{\gamma}_{0k}/y_n) - (\mathbf{N}^\lambda_k y_\lambda / y_n) - \mathbf{N}^n_k]\partial_{(n)} \\ &= \widetilde{\partial}_k - (\mathbf{N}^\lambda_k / L)\widetilde{\partial}_{(\lambda)}. \end{aligned}$$

To prove assertion b), we shall demonstrate that $\widetilde{\partial}_{(n)}$ is perpendicular to each of the vectors $\delta_k, \widetilde{\partial}_{(\varkappa)}$ tangent to IM, which clearly form a basis in

TIM. But,

$$\begin{aligned} G(\delta_k, \widetilde{\partial}_{(n)}) &= G(\delta_k, (y^i/L)\partial_{(i)}) = 0, \\ G(\widetilde{\partial}_{(\varkappa)}, \widetilde{\partial}_{(n)}) &= G(L\partial_{(\varkappa)} - (Ly_\varkappa/y_n)\partial_{(n)}, (y^i/L)\partial_{(i)}) \\ &= y^i g_{\varkappa i} - (y_\varkappa y^i/y_n) g_{ni} \\ &= y_\varkappa - y_\varkappa = 0, \end{aligned}$$

as required. ∎

We set $d\widetilde{x}^k = dx^k$ and $\delta\widetilde{y}^k = (\partial\widetilde{y}^k/\partial y^l)\delta y^l$, where $\delta y^l = dy^l + \mathrm{N}^l_m(x,y)dx^m$. Then $d\widetilde{x}^k, \delta\widetilde{y}^k$ is the dual frame to $\delta_k, \widetilde{\partial}_{(k)}$. Moreover, $d\widetilde{x}^k, \delta\widetilde{y}^\varkappa$ is the dual frame to the frame $\delta_k, \widetilde{\partial}_{(\varkappa)}$ on IM.

Let G be the diagonal lift metric on $\mathring{T}M$. We set

$$g_{(k)(l)} = G(\widetilde{\partial}_{(k)}, \widetilde{\partial}_{(l)}) = \frac{\partial y^i}{\partial \widetilde{y}^k}\frac{\partial y^j}{\partial \widetilde{y}^l} g_{ij}.$$

Then, by Proposition 6.8 b), the metric tensor G on $\mathring{T}M$ and the induced metric tensor $\widehat{G}$ on IM can be written, respectively, as

$$G = g_{kl} d\widetilde{x}^k \otimes d\widetilde{x}^l + g_{(\varkappa)(\lambda)}\delta\widetilde{y}^\varkappa \otimes \delta\widetilde{y}^\lambda + g_{(n)(n)}\delta\widetilde{y}^n \otimes \delta\widetilde{y}^n, \tag{6.13}$$

$$\widehat{G} = g_{kl} d\widetilde{x}^k \otimes d\widetilde{x}^l + g_{(\varkappa)(\lambda)}\delta\widetilde{y}^\varkappa \otimes \delta\widetilde{y}^\lambda. \tag{6.14}$$

From (6.13) and (6.14) it follows that the dual tensors $G^\#$ and $\widehat{G}^\#$ to G and $\widehat{G}$ can be expressed as

$$G^\# = g^{kl}\delta_k \otimes \delta_l + g^{(\varkappa)(\lambda)}\widetilde{\partial}_{(\varkappa)} \otimes \widetilde{\partial}_{(\lambda)} = g^{(n)(n)}\widetilde{\partial}_{(n)} \otimes \widetilde{\partial}_{(n)}, \tag{6.15}$$

$$\widehat{G}^\# = g^{kl}\delta_k \otimes \delta_l + g^{(\varkappa)(\lambda)}\widetilde{\partial}_{(\varkappa)} \otimes \widetilde{\partial}_{(\lambda)},$$

where g^{kl} and $g^{(k)(l)}$ are, respectively, the inverse matrices to g_{kl} and $g_{(k)(l)}$.

By transforming the formulae for the horizontal lift connection ∇ (Definition 6.2) on $\mathring{T}M$ into the frame $\delta_k, \widetilde{\partial}_{(k)}$ and projecting the resulting expressions orthogonally onto TIM, we obtain the following formulae for the induced connection $\widehat{\nabla}$ on IM (cf. [YF80]), where $\widehat{\nabla}$ is called the

K-connection):

$$\begin{aligned}
\widehat{\nabla}_{\delta_k}\delta_l &= \widehat{\Gamma}^m_{kl}\delta_m + \widehat{\Gamma}^{(\mu)}_{kl}\widetilde{\partial}_{(\mu)} \\
&= \mathbf{F}^m_{kl}\delta_m \\
\widehat{\nabla}_{\delta_k}\widetilde{\partial}_{(\lambda)} &= \widehat{\Gamma}^m_{k(\lambda)}\delta_m + \widehat{\Gamma}^{(\mu)}_{k(\lambda)}\widetilde{\partial}_{(\mu)} \\
&= \frac{\partial\widetilde{y}^\mu}{\partial y^m}\left(\delta_k\frac{\partial y^m}{\partial\widetilde{y}^\lambda} + \frac{\partial y^l}{\partial\widetilde{y}^\lambda}\mathbf{F}^m_{kl}\right)\widetilde{\partial}_{(\mu)} \\
\widehat{\nabla}_{\widetilde{\partial}_{(\varkappa)}}\delta_l &= \widehat{\Gamma}^m_{(\varkappa)l}\delta_m + \widehat{\Gamma}^{(\mu)}_{(\varkappa)l}\widetilde{\partial}_{(\mu)} \\
&= \frac{\partial y^m}{\partial\widetilde{y}^\varkappa}\mathbf{C}^m_{kl}\delta_m \\
\widehat{\nabla}_{\widetilde{\partial}_{(\varkappa)}}\widetilde{\partial}_{(\lambda)} &= \widehat{\Gamma}^m_{(\varkappa)(\lambda)}\delta_m + \widehat{\Gamma}^{(\mu)}_{(\varkappa)(\lambda)}\widetilde{\partial}_{(\mu)} \\
&= \frac{\partial\widetilde{y}^\mu}{\partial y^m}\left(\frac{\partial^2 y^m}{\partial\widetilde{y}^\varkappa\partial\widetilde{y}^\lambda} + \frac{\partial y^k}{\partial\widetilde{y}^\varkappa}\frac{\partial y^l}{\partial\widetilde{y}^\lambda}\mathbf{C}^m_{kl}\right)\widetilde{\partial}_{(\mu)}
\end{aligned} \tag{6.16}$$

Since ∇ is a metrical connection on $\mathring{T}M$ with respect to the metric tensor G, it follows that $\widehat{\nabla}$ is a metric connection on IM with respect to the induced metric tensor $\widehat{G}$. In general, the induced connection $\widehat{\nabla}$ has a non-vanishing torsion tensor $\widehat{T}$, which can be obtained by transforming the formulae of Proposition 6.2 for the torsion tensor T of ∇ into the frame $\delta_k, \widetilde{\partial}_{(k)}$ and projecting the resulting expressions orthogonally onto TIM. This procedure yields

$$\begin{aligned}
\widehat{T}(\delta_k,\delta_l) &= \widehat{T}^m_{kl}\delta_m + \widehat{T}^{(\mu)}_{kl}\widetilde{\partial}_{(\mu)} \\
&= \frac{\partial\widetilde{y}^\mu}{\partial y^m}R^m_{kl}\widetilde{\partial}_{(\mu)} \\
\widehat{T}(\delta_k,\widetilde{\partial}_{(\lambda)}) &= \widehat{T}^m_{k(\lambda)}\delta_m + \widehat{T}^{(\mu)}_{k(\lambda)}\widetilde{\partial}_{(\mu)} \\
&= -\frac{\partial y^l}{\partial\widetilde{y}^\lambda}C^m_{lk}\delta_m - \frac{\partial y^l}{\partial\widetilde{y}^\lambda}\frac{\partial\widetilde{y}^\mu}{\partial y^m}P^m_{lk}\widetilde{\partial}_{(\mu)} \\
\widehat{T}(\widetilde{\partial}_{(\varkappa)},\delta_l) &= \widehat{T}^m_{(\varkappa)l}\delta_m + \widehat{T}^{(\mu)}_{(\varkappa)l}\widetilde{\partial}_{(\mu)} \\
&= \frac{\partial y^l}{\partial\widetilde{y}^\lambda}C^m_{kl}\delta_m + \frac{\partial y^l}{\partial\widetilde{y}^\lambda}\frac{\partial\widetilde{y}^\mu}{\partial y^m}P^m_{kl}\widetilde{\partial}_{(\mu)} \\
\widehat{T}(\widetilde{\partial}_{(\varkappa)},\widetilde{\partial}_{(\lambda)}) &= \widehat{T}^m_{(\varkappa)(\lambda)}\delta_m + \widehat{T}^{(\mu)}_{(\varkappa)(\lambda)}\widetilde{\partial}_{(\mu)} \\
&= 0
\end{aligned} \tag{6.17}$$

We are now in a position to consider the Riemannian stochastic development defined by (6.2) in the case of IM equipped with the Riemannian metric $\widehat{G}$ and metric connection $\widehat{\nabla}$. By Theorem 6.3, the projection (X,Y) of this Riemannian stochastic development from the orthonormal frame bundle OIM onto IM is a Riemannian Brownian motion with drift B on IM.

Theorem 6.9. *Let (X,Y) be the above Riemannian Brownian motion with drift B on IM. Then*

a) B can be expressed by the formula in Theorem 6.7,

b) (X,Y) is an indicatrized Finslerian hv-Brownian motion.

Proof. To prove assertion a), we compute the drift B using formula (6.4) from Theorem 6.3 in the case of the indicatrix bundle IM with metric tensor $\widehat{G}$ and metric connection $\widehat{\nabla}$. By (6.15) and (6.17), we have

$$\begin{aligned} B &= \frac{1}{2} g^{km}(\widehat{T}^l_{lm} + \widehat{T}^{(\lambda)}_{(\lambda)m})\delta_k + \frac{1}{2} g^{(\varkappa)(\mu)}(\widehat{T}^l_{l(\mu)} + \widehat{T}^{(\lambda)}_{(\lambda)(\mu)})\widetilde{\partial}_{(\varkappa)} \\ &= \frac{1}{2} g^{km} \frac{\partial y^l}{\partial \widetilde{y}^\lambda} \frac{\partial \widetilde{y}^\lambda}{\partial y^j} P^j_{lm} \delta_k - \frac{1}{2} g^{(\varkappa)(\mu)} \frac{\partial y^m}{\partial \widetilde{y}^\mu} \frac{\partial y^k}{\partial \widetilde{y}^\varkappa} C^l_{ml} \partial_{(k)} \\ &= \frac{1}{2} g^{km}(\delta^l_j - y^l y_j/L^2) P^j_{lm} \delta_k - \frac{1}{2}(g^{km} - y^k y^m/L^2) C^l_{ml} \partial_{(k)}. \end{aligned} \tag{6.18}$$

The last equality holds, since

$$\begin{aligned} \frac{\partial y^i}{\partial \widetilde{y}^\varkappa} \frac{\partial \widetilde{y}^\varkappa}{\partial y_j} &= \frac{\partial y^i}{\partial \widetilde{y}^k} \frac{\partial \widetilde{y}^k}{\partial y_j} - \frac{\partial y^i}{\partial \widetilde{y}^n} \frac{\partial \widetilde{y}^n}{\partial y_j} \\ &= \delta^i_j - y^i y_j / L^2, \end{aligned} \tag{6.19}$$

$$\begin{aligned} g^{(\varkappa)(\mu)} \frac{\partial y^i}{\partial \widetilde{y}^\varkappa} \frac{\partial y^j}{\partial \widetilde{y}^\mu} &= g^{(k)(m)} \frac{\partial y^i}{\partial \widetilde{y}^k} \frac{\partial y^j}{\partial \widetilde{y}^m} - g^{(n)(n)} \frac{\partial y^i}{\partial \widetilde{y}^n} \frac{\partial y^j}{\partial \widetilde{y}^n} \\ &= g^{ij} - y^i y^j / L^2 \end{aligned} \tag{6.20}$$

Because $\mathbf{N}^j_m(x,y)$ is homogeneous of degree one in y, it follows by the Euler theorem that $y^l \partial_{(l)} \mathbf{N}^j_m(x,y) = \mathbf{N}^j_m(x,y)$, which implies that

$$y^l y_j P^j_{lm} = y^l y_j (\partial_{(l)} \mathbf{N}^j_m - \mathbf{F}^j_{ml}) = y_j \mathbf{N}^j_m - y_j \mathbf{N}^j_m = 0.$$

We also have $y^m \mathbf{C}^l_{ml}(x,y) = 0$. Since $L(x,y) = 1$ on IM, the formula for B therefore follows from (6.18).

To prove assertion b), we compute the generator D of (X,Y) using formula (6.3) from Theorem 6.3 in the case of the indicatrix bundle IM with metric tensor $\widehat{G}$ and metric connection $\widehat{\nabla}$. Writing the first equality in (6.3) in terms of the frame $\delta_k, \widetilde{\partial}_{(k)}$, by (6.15), (6.16), (6.19), (6.20), and the identity $y_i \mathrm{C}^i_{kl}(x,y) = 0$, we have

$$\begin{aligned}
D &= \frac{1}{2} g^{kl}(\delta_k \delta_l - \widehat{\Gamma}^m_{kl}\delta_m - \widehat{\Gamma}^{(\mu)}_{kl}\widetilde{\partial}_{(\mu)}) \\
&\quad + \frac{1}{2} g^{(\varkappa)(\lambda)}(\widetilde{\partial}_{(\varkappa)}\widetilde{\partial}_{(\lambda)} - \widehat{\Gamma}^m_{(\varkappa)(\lambda)}\delta_m - \widehat{\Gamma}^{(\mu)}_{(\varkappa)(\lambda)}\widetilde{\partial}_{(\mu)}) \\
&= \frac{1}{2} g^{kl}(\delta_k \delta_l - \mathrm{F}^m_{kl}\delta_m) + \frac{1}{2} g^{(\varkappa)(\lambda)}\Big(\frac{\partial y^k}{\partial \widetilde{y}^\varkappa}\frac{\partial y^l}{\partial \widetilde{y}^\lambda}\partial_{(k)}\partial_{(l)} + \frac{\partial^2 y^m}{\partial \widetilde{y}^\varkappa \partial \widetilde{y}^\lambda}\partial_{(m)}\partial_m \\
&\quad - \frac{\partial^2 y^i}{\partial \widetilde{y}^\varkappa \partial \widetilde{y}^\lambda}\frac{\partial \widetilde{y}^\mu}{\partial y^i}\frac{\partial y^m}{\partial \widetilde{y}^\mu}\partial_{(m)} - \frac{\partial y^k}{\partial \widetilde{y}^\varkappa}\frac{\partial y^l}{\partial \widetilde{y}^l}\frac{\partial \widetilde{y}^\mu}{\partial y^i}\frac{\partial y^m}{\partial \widetilde{y}^\mu}\mathrm{C}^i_{kl}\partial_{(m)}\Big) \\
&= \frac{1}{2} g^{kl}(\delta_k \delta_l - \mathrm{F}^m_{kl}\delta_m) + \frac{1}{2}(g^{kl} - y^k y^l/L^2)(\partial_{(k)}\partial_{(l)} - \mathrm{C}^m_{kl}\partial_{(m)}) \\
&\quad + \frac{1}{2} g^{(\varkappa)(\lambda)}\frac{\partial^2 y^i}{\partial \widetilde{y}^\varkappa \partial \widetilde{y}^\lambda} y_i (y^m/L^2)\partial_{(m)}.
\end{aligned}$$

To evaluate the last term in the above formula, we compute the following second-order partial derivative, the result being equal to zero, since $L(x,y) = \widetilde{y}^n$ is independent of $\widetilde{y}^\lambda$:

$$\begin{aligned}
0 &= \frac{\partial^2 L^2(x,y)}{\partial \widetilde{y}^\varkappa \partial \widetilde{y}^\lambda} = 2\frac{\partial}{\partial \widetilde{y}^\varkappa}\Big(\frac{\partial y^i}{\partial \widetilde{y}^\lambda} y_i\Big) \\
&= 2\frac{\partial^2 y^i}{\partial \widetilde{y}^\varkappa \partial \widetilde{y}^\lambda} y_i + 2\frac{\partial y^i}{\partial \widetilde{y}^\lambda}\frac{\partial y^j}{\partial \widetilde{y}^\varkappa} g_{ij}.
\end{aligned}$$

Thus,

$$\begin{aligned}
g^{(\varkappa)(\lambda)}\frac{\partial^2 y^i}{\partial \widetilde{y}^\varkappa \partial \widetilde{y}^\lambda} y_i &= -g^{(\varkappa)(\lambda)}\frac{\partial y^i}{\partial \widetilde{y}^\lambda}\frac{\partial y^j}{\partial \widetilde{y}^\varkappa} g_{ij} \\
&= -(g^{ij} - y^i y^j/L^2) g_{ij} \\
&= -(n-1).
\end{aligned}$$

Since $L(x,y) = 1$ on IM, it follows that D is given by formula (6.9), i.e., (X,Y) is an indicatrized Finslerian hv-Brownian motion. ∎

Proof of Theorem 6.7. If follows from Theorem 6.9 that the generator D of an indicatrized hv-Brownian motion given by (6.9) is equal to $\frac{1}{2}\Delta + B$, where Δ is the Laplace-Beltrami operator on IM equipped with the Riemannian metric $\hat{G}$ and where B is the vector field in IM defined by (6.11). This proves Theorem 6.7. ∎

Appendix A

Diffusion and Laplacian on the Base Space

Our aim in this appendix is to present a construction of the Laplace operator on the base space M of a Finsler manifold rather than on $\mathring{T}M$ or IM, as in the main body of the book. The operator will be applied to define harmonic forms and to obtain a Hodge decomposition theorem for Finsler spaces. The vehicle we employ in our construction is diffusion theory, in particular, a result which can be understood as the Central Limit Theorem for geodesic random walks. This extends Pinsky's results on isotropic random walks and their limit on a Riemannian manifold [Pin78], [Pin92]. Diffusion theory makes it possible to resolve the well-known difficulties inherent in studying harmonic forms on a Finsler manifold. The Laplacian emerging from diffusion on the base space M provides a natural definition of harmonic forms as those that realize heat equilibria.

We shall denote our Laplacian by Δ_{AZ} to distinguish it from that proposed by Bao and Lackey in their ground-breaking paper [BL96]. We shall refer to the latter as the BL-Laplacian and denote it by Δ_{BL}. It was constructed by considering the indicatrix bundle IM as a Riemannian manifold with the diagonal lift metric G (for the definition, see Chapter 6), which gives rise to a Hodge star $*$ on IM and thus an inner product

$$\langle \eta, \theta \rangle = \int_{IM} \frac{(-1)^{n-1}}{\mathrm{Vol}(IM_x)} \eta \wedge *\theta$$

of differential forms η, θ on IM. This can also be regarded as an inner product of differential forms η, θ on M, since any such form can be identified with a form on IM (depending only on $x \in M$). Taking d_M to be the exterior differential on M and d_M^* the adjoint to d_M in the sense that

$$\langle d_M^* \eta, \theta \rangle = \langle \eta, d_M \theta \rangle$$

for any differential forms η, θ on M, the *BL-Laplacian* Δ_{BL} on M can be defined by

$$\Delta_{\mathrm{BL}} = d_M^* d_M + d_M d_M^*.$$

Because an inner product has been introduced on the elliptic complex of differential forms on M, a Hodge decomposition theorem follows by a general result about elliptic complexes in Wells [Wel80], Chap. 4, Theorem 5.2.

Is any of the two Laplacians Δ_{BL} and Δ_{AZ} more natural than the other? Or could perhaps yet another candidate on which to base Finslerian Hodge Theory be revealed? After all, as far as the Hodge decomposition theorem is concerned, it only matters that there should be a well-defined inner product on differential forms, which leaves plenty of room for choice. Our Laplacian Δ_{AZ} emerging from diffusion theory appears to be the most natural one in that the resulting harmonic forms realize heat equilibria, that is, they are invariant under the heat flow generated by diffusion on M. This completes the first step of the Milgram-Rosenbloom program for Finsler Hodge theory.

The construction of Δ_{AZ} will proceed as follows. First we shall introduce a geodesic random walk on a Finsler manifold M extending Pinsky's concept of an isotropic random walk on a Riemannian manifold [Pin78], [Pin92]. Then a limit will be taken under suitable scaling as in the Central Limit Theorem, producing a diffusion generator on M, which is obviously a second-order operator. The quadratic form of this operator will give rise to an inner product of differential forms on M, leading eventually to the definition of our Laplacian Δ_{AZ} in a similar way as for Δ_{BL} above. Harmonic forms are then introduced and a Hodge decomposition obtained again by a general result about elliptic complexes in [Wel80].

Throughout this appendix M will be a without boundary compact Finsler manifold with smooth metric function $L : \overset{\circ}{T}M \to \mathbb{R}$ positive definite metric tensor $g : \overset{\circ}{T}M \to T_2 M$. This ensures that for each $x \in M$ the indicatrix IM_x is a smooth compact submanifold in $\overset{\circ}{T}M$. In addition, we shall assume that L homogeneous in y, that is,

$$L(x, ky) = |k| L(x, y),$$

and not just positive homogeneous.

A.1 Finslerian Isotropic Transport Process

To define this process we first introduce the *geodesic flow* on a Finsler manifold. This is a deterministic motion $\xi_t(x,y)$ obtained by following the unit speed geodesic with initial position $\xi_0(x,y) = x \in M$ and initial velocity $\dot{\xi}_0(x,y) = y \in IM_x$. It satisfies the geodesic equation

$$\ddot{\xi}_t^i + \gamma_{jk}^i(\xi_t, \dot{\xi}_t)\dot{\xi}_t^j\dot{\xi}_t^k = 0,$$

where $\gamma_{jk}^i(x,y)$ are the Finslerian formal Christoffel symbols (1.55). If we put

$$(T_t^0 f)(x,y) = f(\xi_t(x,y), \dot{\xi}_t(x,y))$$

for any function $f \in C^\infty(IM)$, then $T_{s+t}^0 = T_s^0 T_t^0$ and

$$u(t,x,y) = \left(T_t^0 f\right)(x,y)$$

satisfies the p.d.e. $\partial_t u = Zu$, where the first-order operator

$$Z = y^i\partial_i - \gamma_{jk}^i(x,y)y^jy^k\dot{\partial}_i = y^i\delta_i$$

is known as the *geodesic flow field* or the *canonical horizontal vector field* on IM. It is the infinitesimal operator of the semigroup T_t^0, $t \geq 0$.

Next, for any $x \in M$ we consider the indicatrix IM_x as a submanifold of IM with the diagonal lift metric G induced from IM. This means simply that $g_{ij}(x,y)$ considered as a function of y with x fixed is taken as a metric tensor on $\mathring{T}M_x$ and then it is restricted to $IM_x \subset \mathring{T}M_x$. The indicatrix IM_x equipped with this metric is a compact Riemannian manifold. We denote by ω_x the volume measure of IM_x in this metric and assume that it is normalized so that

$$\int_{IM_x} \omega_x(dy) = 1.$$

The *canonical projection operator* $\Pi : C^\infty(IM) \to C^\infty(M)$ is then defined by

$$(\Pi f)(x) = \int_{IM_x} f(x,y)\, d\omega_x(y).$$

Let us take a sequence of independent random variables $e_1, e_2, \ldots$ with exponential distribution

$$P\{e_n > t\} = e^{-t}, \quad t \geq 0,$$

on some probability space (Ω, Σ, P). Then $t_n = e_1 + \cdots + e_n$ are the jump times of a Poisson process with unit rate. We also put $t_0 = 0$ for convenience.

The isotropic transport process can be defined by induction. For any given starting point $x \in M$ and starting velocity $y \in IM_x$ we put $x_0 = x$ and $y_0 = y$. Assuming that $x_i \in M$ and $y_i \in IM_{x_i}$ have been constructed for $i = 0, \ldots, n-1$, we put

$$x_n = \xi_{e_n}(x_{n-1}, y_{n-1})$$

and require that the conditional distribution of $y_n \in IM_{x_i}$ given $e_1, \ldots, e_n$ and $y_1, \ldots, y_{n-1}$ should be uniform on IM_{x_n} under the normalized volume measure ω_{x_n}, that is,

$$E\{f(x_n, y_n)|e_1, \ldots, e_n, y_1, \ldots, y_{n-1}\} = (\Pi f)(x_n)$$

for any $f \in C^\infty(IM_{x_n})$. Finally, we put

$$X_t = \xi_{t-t_n}(x_n, y_n) \quad \text{and} \quad Y_t = \dot{X}_t \quad \text{for } t \in [t_n, t_{n+1}),\ n = 0, 1, 2, \ldots .$$

In this way we obtain a Markov process (X_t, Y_t) on the indicatrix bundle IM defined for all $t \geq 0$, called the *isotropic transport process.* The sample paths of X_t are continuous piecewise geodesic with jumps of the tangent vector Y_t at the jump times t_n of the Poisson process.

The infinitesimal operator L of (X_t, Y_t) can be computed by observing that on $\{t < e_1\}$, which has probability e^{-t}, the process moves along the original geodesic $\xi_t(x, y)$, and on $\{t \geq e_1\}$, which has probability $1 - e^{-t}$, a new direction y_1 is chosen according to the measure ω_{x_1} at time t_1. For any $f \in C^\infty(IM)$ we have

$$f(X_t, Y_t) = f(\xi_t(x, y), \dot{\xi}_t(x, y)) = f(x, y) + t(Zf)(x, y) + o(t)$$

on $\{t < e_1\}$ and

$$\begin{aligned} f(X_t, Y_t) &= f(x_1, y_1) + o(1) \\ (\Pi f)(x_1) &= (\Pi f)(x) + o(1) \end{aligned}$$

on $\{t \geq e_1\}$. Using the fact that $E\{f(x_1, y_1)|e_1\} = (\Pi f)(x_1)$, we obtain

$$\begin{aligned} Ef(X_t, Y_t) &= \int_{\{t<e_1\}} f(X_t, Y_t)dP + \int_{\{t\geq e_1\}} f(X_t, Y_t)dP \\ &= e^{-t}[f(x,y) + t(Zf)(x,y)] + \int_{\{t\geq e_1\}} f(x_1, y_1)dP + o(t) \\ &= e^{-t}[f(x,y) + t(Zf)(x,y)] + \int_{\{t\geq e_1\}} (\Pi f)(x_1)dP + o(t) \\ &= e^{-t}[f(x,y) + t(Zf)(x,y)] + (1 - e^{-t})(\Pi f)(x) + o(t) \\ &= f(x,y) + t[(Zf)(x,y) + (\Pi f)(x) - f(x,y)] + o(t). \end{aligned}$$

It follows that

$$Lf = \lim_{t\downarrow 0} \frac{1}{t}(f \circ T_t - f) = Zf + \Pi f - f.$$

Here T_t is the semigroup of (X_t, Y_t),

$$(T_t f)(x,y) = Ef(X_t, Y_t)$$

for any $f \in C^\infty(IM)$.

A.2 Central Limit Theorem

We shall scale the isotropic transport process in such a way that the mean time and distance travelled between consecutive jumps of direction will be ε^2 and ε, respectively, where $\varepsilon > 0$. In other words, the scaled process X_t^ε will be piecewise geodesic with speed $1/\varepsilon$ and jumps of direction at the jump times of a Poisson process with rate $1/\varepsilon^2$. The velocity part of the process will be defined by $Y_t^\varepsilon = \varepsilon \dot{X}_t^\varepsilon$ to ensure that the joint process $(X_t^\varepsilon, Y_t^\varepsilon)$ lives on the unit-radius indicatrix bundle IM. The infinitesimal operator of the scaled process $(X_t^\varepsilon, Y_t^\varepsilon)$ is then found to be

$$L^\varepsilon = (1/\varepsilon)Z + (1/\varepsilon^2)(\Pi - 1).$$

We claim that X_t^ε tends weakly to a diffusion process on M as $\varepsilon \downarrow 0$.

First we observe that $\Pi^2 = \Pi$ and $\Pi Z \Pi = 0$. The latter equality depends on the fact that any linear function of y on IM_x has integral zero with respect to the normalized volume measure ω_x. This will be so if the indicatrix IM_x is symmetric about the origin in the tangent space TM_x, which is the case if the Finsler metric function L is homogeneous in y (rather than just positive homogeneous). Using these two equalities, we obtain the identity

$$L^\varepsilon(1+\varepsilon Z+\varepsilon^2 Z^2)\Pi = \Pi Z^2\Pi + \varepsilon Z^3\Pi.$$

This implies the following lemma.

Lemma A.1. *For any $f \in C^\infty(M)$ put*

$$f^\varepsilon = (1+\varepsilon Z+\varepsilon^2 Z^2)f \in C^\infty(IM).$$

Then

$$\lim_{\varepsilon\downarrow 0} f^\varepsilon = f \quad \text{and} \quad \lim_{\varepsilon\downarrow 0} L^\varepsilon f^\varepsilon = \Pi Z^2 f.$$

The operator ΠZ^2 on the right-hand side is the generator of a diffusion X_t on M. The theory of convergence of semigroups, in particular, the limit theorem for semigroups in application to random evolution proved by Kurtz in [Kur73] can now be applied to show that

$$\lim_{\varepsilon\downarrow 0} e^{tL^\varepsilon} f = e^{t\Pi Z^2} f$$

for all $f \in C^\infty(M)$, which, in turn, implies that X_t^ε converges weakly to X_t in the sense that

$$\lim_{\varepsilon\downarrow 0} E\{f(X_t^\varepsilon)\} = E\{f(X_t)\}$$

for every $f \in C(M)$. All technical assumptions leading to these results are trivially satisfied if M is compact.

The diffusion generator ΠZ^2 can be expressed as

$$(\Pi Z^2 f)(x) = \int_{IM_x} y^i y^j \left(\partial_i\partial_j - \gamma^k_{ij}(x,y)\partial_k\right) f(x)\, d\omega_x(y)$$

for any $f \in C^\infty(M)$.

This limiting argument is based Pinsky's approach to isotropic transport and diffusion on Riemannian manifolds, see [Pin78], [Pin92], where further technical details can also be found. In particular, if M is a Riemannian manifold, then

$$g^{ij}(x) = (\dim M)\int_{IM_x} y^i y^j d\omega_x(y)$$

and

$$\Delta = (\dim M)\,\Pi Z^2,$$

the Laplace-Beltrami operator, so $X_{t/\dim M}$ is a Brownian motion on M. This has been extended above to the case of Finsler manifolds.

A.3 Laplacian, Harmonic Forms and Hodge Decomposition

The diffusion generator ΠZ^2 is a second-order differential operator with leading quadratic form

$$H^{ij}(x) = \int_{IM_x} y^i y^j \, d\omega_x(y),$$

which is clearly non-degenerate and positive definite. When M is Riemannian, it reduces to the metric tensor $g^{ij}(x) = (\dim M)\, H^{ij}(x)$. Similarly, we define

$$H^{i_1 j_1 \ldots i_p j_p}(x) = \int_{IM_x} y^{i_1} y^{j_1} \ldots y^{i_p} y^{j_p} \, d\omega_x(y),$$

which can be written briefly as

$$H^{IJ}(x) = \int_{IM_x} y^I y^J \, d\omega_x(y)$$

for multi-indices I and J. Because

$$H^{IJ}(x) u_I u_J = \int_{IM_x} \left(y^I u_I\right)^2 d\omega_x(y) > 0 \tag{A.1}$$

for any p-form $u \neq 0$, it follows that $H^{IJ}(x)$ defines an inner product on the exterior algebra $\Lambda^p_x M$, namely,

$$(\eta|\theta) = \int_M \eta_I \theta_J H^{IJ} \sqrt{H} \, dx,$$

where

$$\sqrt{H} = \int_{IM_x} \sqrt{g} \, d\omega_x(y), \qquad \sqrt{g} = \sqrt{\det(g_{ij})}.$$

It is easily seen that $\sqrt{H}$ is a tensor density of weight 1 and H^{IJ} is a contravariant tensor of rank $2p$ on M, so the product $(\eta|\theta)$ is well defined. It follows from (A.1)) that H^{IJ} is invertible as a map from p-forms to p-forms. The inverse will be denoted by $H_{IJ} = H_{i_1 j_1 \ldots i_p j_p}$.

We denote by d the exterior differential on M and by δ the codifferential defined to be the adjoint to d in the above inner product,

$$(d\eta|\theta) = (\eta|\delta\theta)$$

for any $\eta \in \Lambda^{p-1} M$ and $\theta \in \Lambda^p M$.

Definition A.1. The AZ-Laplacian on the base manifold M is then defined by

$$\Delta_{\mathrm{AZ}} = \delta d + d\delta.$$

Proposition A.2. *Let M be a without boundary compact Finsler manifold with smooth metric function $L(x, y)$ homogeneous (rather than just positive homogeneous) in y and such that the Finsler metric tensor metric tensor $g_{ij}(x, y)$ is positive definite. Then*

$$\Delta_{AZ}\eta = 0 \Leftrightarrow d\eta = \delta\eta = 0 \tag{A.2}$$

for any $\eta \in \Lambda^p M$.

Proof. Indeed, if $\Delta_{\mathrm{AZ}}\eta = 0$, then

$$0 = (\Delta_{\mathrm{AZ}}\eta|\eta) = (\delta d\eta|\eta) + (d\delta\eta|\eta) = (d\eta|d\eta) + (\delta|\delta\eta)\,,$$

so $d\eta = \delta\eta = 0$. ∎

Definition A.2. A form $\eta \in \Lambda^p M$ satisfying either side of the equivalence (A.22) will be called *harmonic* and the space of such forms will be denoted by $\mathcal{H}^p M$.

Finally, we are in a position to state the following Hodge decomposition theorem based on our Laplacian Δ_{AZ}.

Theorem A.3. *Let M be a Finsler manifold satisfying the assumptions of Proposition A.2, with harmonic functions defined by means of the AZ-Laplacian Δ_{AZ}. Then*

(a) *Each cohomology class $H^p M$ contains a unique harmonic representative.*

(b) *$\mathcal{H}^p M$ is finite dimensional, its dimension being equal to the pth Betti number of M.*

(c) *$\Lambda^p M = \mathcal{H}^p M \oplus d\Lambda^{p-1} M \oplus \delta\Lambda^{p+1} M$, the three spaces on the right-hand side being mutually orthogonal in L_2.*

Proof. The proof is exactly the same as in Bao and Lackey's paper [BL96]. It follows directly from the general result about elliptic complexes in Wells [Wel80], Chap. 4, Theorem 5.2, or can be proved by the elegant and simple argument presented in [BL96]. Rather than repeating this word-by-word, we refer the reader to the sources. ∎

Appendix B

Two-Dimensional Constant Berwald Spaces

B.1 Berwald's Famous Theorem

We consider an n-dimensional Finsler space $F^n = \left(M^n, L(x,y)\right) :\ (x^i)$ local coordinates and (y^i) tangent vectors; therefore,

$$
\begin{aligned}
F &= L^2/2, \quad g_{ij} = \dot{\partial}_i \dot{\partial}_j F, \quad (g^{ij}) = (g_{ij})^{-1}, \\
\gamma^i_{jk} &= g^{ir}(\partial_k g_{rj} + \partial_j g_{rk} - \partial_r g_{jk})/2, \\
2G^i &= g^{ij}\{y^r(\dot{\partial}_j \partial_r F) - \partial_j F\} = \gamma^i_{jk} y^j y^k, \\
G^i_j &= \dot{\partial}_j G^i, \quad G^i_{jk} = \dot{\partial}_k G^i_j, \quad 2G^i = G^i_{jk} y^j y^k, \\
& \qquad \text{with the geodesics:} \quad \frac{d^2 x^i}{ds^2} + 2G^i\left(x, \frac{dx}{ds}\right) = 0.
\end{aligned}
$$

F^n is called a *Berwald space*, if G^i_{jk} depend on $x = (x^i)$ alone. We shall find all the Finsler spaces with constant G^i_{jk} from the set of Berwald spaces of dimension $n = 2$.

F^n is called a *locally Minkowski space*, if L depends on $y = (y^i)$ alone for some choice of $x = (x^i)$. Then G^i_{jk} of F^n vanish, hence a locally Minkowski space is, of course, a kind of Berwald space.

We have the following celebrated theorem on two-dimensional Berwald spaces which are not locally Minkowski.

All the two-dimensional Berwald spaces which are not locally Minkowski are divided into the following three classes according to the magnitude of the main scalar I:

(1) $B^2(1) : I^2 < 4$,

$$L = \sqrt{(a^1)^2 + (a^2)^2} \exp\{J \tan^{-1}(a^1/a^2)\},$$
$$J = I/\sqrt{4 - I^2},$$

(2) $B^2(2) : I^2 = 4$,

$$L = |a^1| \exp(\varepsilon a^2/a^1), \quad \varepsilon = I/2 = \pm 1,$$

(3) $B^2(r, s) : I^2 > 4$,

$$L = |(a^1)^r (a^2)^s|, \quad r + s = 1,$$

where $a^\alpha(x, y) = a^\alpha_1(x)y^1 + a^\alpha_2(x)y^2$, $\alpha = 1, 2$, *are independent* 1-*forms.* This is the celebrated Theorem of Berwald, [Mat94].

Theorem B.1. $2G^i(x, y)$ *of the spaces belonging to* $B^2(1)$, $B^2(2)$ *and* $B^2(r, s)$ *are written in the form*

$$2G^i(x, y) = \{\Gamma^i_{jk}(x) + T^{\,i}_{\cdot jk}(x, y)\} y^j y^k,$$

where we put

$$\Gamma^i_{jk}(x) = b^i_\alpha \cdot \partial_k a^\alpha_j, \quad (b^i_\alpha) = (a^\alpha_i)^{-1},$$
$$T^i_{jk}(x) = \Gamma^i_{jk} - \Gamma^i_{kj}, \quad T^{\,i}_{\cdot jk} = g^{ir} g_{js} T^s_{rk}.$$

Putting $d =$ det $(a^\alpha_i) \neq 0$ *and* $A^\alpha = \partial_2 a^\alpha_1 - \partial_1 a^\alpha_2$, *we have*

$$(1)\ B^2(1) : \begin{cases} T^1_{\cdot 00} &= (A^\alpha b_\alpha)(a^\beta_2 b_\beta)/d^2(1 + J^2), \\ T^2_{\cdot 00} &= -(A^\alpha b_\alpha)(a^\beta_1 b_\beta)/d^2(1 + J^2), \end{cases}$$

$$b_1 = a^1 + J a^2, \quad b_2 = a^2 - J a^1,$$

$$(2)\ B^2(2) : \begin{cases} T^1_{\cdot 00} &= (A^\alpha c_\alpha)(a^\beta_2 c_\beta)/d^2, \\ T^2_{\cdot 00} &= -(A^\alpha c_\alpha)(a^\beta_1 c_\beta)/d^2, \end{cases}$$

$$c_1 = a^1 - \varepsilon a^2, \quad c_2 = \varepsilon a^1,$$

$$(3)\ B^2(r, s) : \begin{cases} T^1_{\cdot 00} &= -T(s a^2_2 a^1 + r a^1_2 a^2)/rsd^2, \\ T^2_{\cdot 00} &= T(s a^2_1 a^1 + r a^1_1 a^2)/rsd^2, \end{cases}$$

$$T = rA^1 a^2 + sA^2 a^1.$$

B.2 Standard Coordinate Representation

Let $1/a(x)$ and $1/b(x)$ be *integrating factors* of a^1 and a^2, respectively. That is, we have the functions $\bar{x}^1(x)$ and $\bar{x}^2(x)$ such that

$$\begin{aligned} a^1(x, dx) &= a^1_1(x)dx^1 + a^1_2(x)dx^2 = a(x)d\bar{x}^1, \\ a^2(x, dx) &= a^2_1(x)dx^1 + a^2_2(x)dx^2 = b(x)d\bar{x}^2. \end{aligned}$$

The pair $(\bar{x}^1, \bar{x}^2)$ may be regarded as a new local coordinate system, because the Jacobian $\partial(\bar{x}^1, \bar{x}^2)/\partial(x^1, x^2) = d/ab$ does not vanish. Such a coordinate system is called a *standard coordinate system of a two-dimensional Finsler space with 1-form metric.* It is noted that for a standard (x^1, x^2) coordinate system, (hx^1, kx^2) with non-zero constant h, k is also standard.

We shall write the expression of $G^i_{jk}(x)$ of Finsler spaces belonging to the classes $B^2(1)$, $B^2(2)$ and $B^2(r, s)$ respectively, in a *standard coordinate system*: $(x, y) \equiv (x^1, x^2)$ and $(\dot{x}, \dot{y}) \equiv (y^1, y^2)$.

(1) $\boldsymbol{B^2(1)}$:

$$\begin{aligned} ab(1+J^2)G^1_{11} &= ba_x(1+J^2) - Ja(a_y + Jb_x), \\ a(1+J^2)G^1_{12} &= a_y + Jb_x, \\ a^2(1+J^2)G^1_{22} &= b(Ja_y - b_x), \\ b^2(1+J^2)G^2_{11} &= -a(a_y + Jb_x), \\ b(1+J^2)G^2_{12} &= b_x - Ja_y, \\ ab(1+J^2)G^2_{22} &= ab_y(1+J^2) - Jb(Ja_y - b_x), \end{aligned} \tag{B.1}$$

(2) $\boldsymbol{B^2(2)}$: The surviving (i.e. non-zero) G^i_{jk} are

$$\begin{aligned} &G^1_{11} = a_x/a + \varepsilon a_y/b - b_x/b, \quad G^2_{12} = \varepsilon a_y/b, \\ &G^2_{11} = -(a/b)(a_y/b - \varepsilon b_x/b), \quad G^2_{22} = b_y/b - a_y/a. \end{aligned} \tag{B.2}$$

(3) $\boldsymbol{B^2(r, s)}$: The surviving G^i_{jk} are

$$\begin{aligned} &G^1_{11} = (rA_x + sB_x)/r, \quad G^2_{22} = (rA_y + sB_y)/s, \\ &A = \log|a|, \qquad\qquad B = \log|b|. \end{aligned} \tag{B.3}$$

B.3 $B^2(1)$ with Constant G^i_{jk}

We shall find the Finsler spaces belonging to $B^2(1)$ which have constant G^i_{jk} in a standard coordinate system.

From the equations (B.1), we have generally

$$a^2G^1_{22} + b^2G^2_{12} = a^2G^1_{12} + b^2G^2_{11} = 0. \tag{B.4}$$

Now, assume that all the G^i_{jk} are constant. Then (B.4) leads us to the following two cases:

(1°) G^2_{12} or $G^2_{11} \neq 0$. Then $c = b/a$ is constant.

(2°) $G^2_{12} = G^2_{11} = 0$. Then $G^1_{22} = G^1_{12} = 0$.

First, we deal with the case (1°). Since (x, cy) may be regarded as the new standard coordinate system (x, y), we may take $c = 1$. Consequently, (B.1) reduces to

$$\begin{aligned} G^1_{11} &= G^2_{12} = -G^1_{22} = (a_x - Ja_y)/a(1+J^2), \\ G^1_{12} &= G^2_{22} = -G^2_{11} = (a_y + Ja_x)/a(1+J^2), \end{aligned} \tag{B.5}$$

which imply

$$a_x/a = G^1_{11} + JG^2_{22}, \quad a_y/a = G^2_{22} - JG^1_{11}.$$

Thus $a_x/a = a_1$ and $a_y/a = a_2$ must be constant, hence we have $a = \exp(a_1x + a_2y + a_0)$ with another constant a_0.

Proposition B.2. *A Finsler space which belongs to $B^2(1)$ and has constant G^i_{jk} in standard coordinates is such that*

$$L(x, y; \dot{x}, \dot{y}) = \sqrt{\dot{x}^2 + \dot{y}^2}\, \exp\{a_1x + a_2y + a_0 + J\tan^{-1}(\dot{x}/\dot{y})\},$$

where a_1, a_2 and a_0 are constant, and the dots denote d/ds. Also,

$$\begin{aligned} G^1_{11} &= G^1_{12} = -G^1_{22} = (a_1 - Ja_2)/(1+J^2) = c_1, \\ G^2_{22} &= G^1_{12} = -G^2_{11} = (a_2 + Ja_1)/(1+J^2) = c_2, \end{aligned}$$

and the geodesics are

$$\begin{cases} \ddot{x} + c_1\dot{x}^2 + 2c_2\dot{x}\dot{y} - c_1\dot{y}^2 &= 0, \\ \ddot{y} - c_2\dot{x}^2 + 2c_1\dot{x}\dot{y} + c_2\dot{y}^2 &= 0, \end{cases}$$

or

$$y'' = \{(y')^2 + 1\}(c_2 - c_1y'), \quad y' = dy/dx.$$

Remark B.1. It is obvious that a_0 may be reduced to zero by the homothetic transformation. Thus we obtain *P.L. Antonelli's metric* where $\alpha_i = a_i/(1+J^2)$, [Ant90], [Ant91].

The case (2°) leads us only to locally Minkowski spaces. In fact, $G^1_{12} = G^2_{12} = 0$ yield $a_y = b_x = 0$, hence $a = a(x)$ and $b = b(y)$. Then

$$G^1_{11} = a_x/a = a_1, \quad G^2_{22} = b_y/b = b_1,$$

hence we have $a = \exp(a_1 x + a_0)$ and $b = \exp(b_1 y + b_0)$ with constant a_0 and b_0. If $a_1 b_1 \neq 0$, then we put

$$a_1 \bar{x} = \exp(a_1 x + a_0), \quad b_1 \bar{y} = \exp(b_1 y + b_0).$$

If $a_1 = 0$, then we put $\bar{x} = (\exp a_0)x$, then it is easy to show that L does not contain either $\bar{x}$ or $\bar{y}$.

B.4 Class $B^2(2)$ with Constant G^i_{jk}

From (2) we can solve for a_x, a_y, b_x and b_y as follows:

$$\begin{cases} a_x = aG^1_{11} + \varepsilon b G^2_{11}, & a_y = \varepsilon b G^2_{12}, \\ b_x = b(\varepsilon b G^2_{11}/a + G^2_{12}), & b_y = b(G^2_{22} + \varepsilon b G^2_{12}/a). \end{cases} \tag{B.6}$$

Or, we put $b = ca$ and obtain

$$\begin{cases} a_x/a = G^1_{11} + \varepsilon c G^2_{11}, & a_y/a = \varepsilon c G^2_{12}, \\ c_x/c = G^2_{12} - G^1_{11}, & c_y/c = G^2_{22}. \end{cases} \tag{B.7}$$

Suppose that all the G^i_{jk} be constant. Putting

$$c_1 = G^2_{12} - G^1_{11}, \quad c_2 = G^2_{22}, \tag{B.8}$$

two equations of (B.7) yield $c = \exp(c_1 x + c_2 y + c_0)$ with another constant c_0. Similarly as in Section B.3, c_0 may be taken $c_0 = 0$. Hence

$$c = \exp(c_1 x + c_2 y). \tag{B.9}$$

Next, if we put $A = \log|a|$, the remaining two equations of (B.7) are written as

$$A_x = G^1_{11} + \varepsilon c G^2_{11}, \quad A_y = \varepsilon c G^2_{12}, \tag{B.10}$$

which imply $c_y G^2_{11} = c_x G^2_{12}$, that is,

$$c_2 G^2_{11} = c_1 G^2_{12}, \tag{B.11}$$

which is a necessary condition for our assumption. Then (B.10) yields

$$A = G^1_{11}x + \varepsilon G^2_{11}\int c dx + f(y) = \varepsilon G^2_{12}\int c dy + g(x).$$

Therefore, we have

(1°) $c_1 \neq 0: \quad A = G^1_{11}x + \varepsilon G^2_{11}(c/c_1) + c_0,$

where c_0 is another constant.

(2°) $c_1 = c_2 = 0: \quad A = (G^1_{11} + \varepsilon c G^2_{11})x + \varepsilon c G^2_{12} y + c_0,$

where c_0 is a constant. On the other hand, we have $c =$ exp c_0; this constant c may be taken as $c = 1$ because $a = e^A$ and $b = ce^A$, as in Section B.3.

Proposition B.3. *A Finsler space which belongs to $B^2(2)$ and has constant G^i_{jk} in standard coordinates is such that*

(1°) $L = |\dot{x}| \exp(A + \varepsilon c\dot{y}/\dot{x}), \quad A = G^1_{11}x + \varepsilon G^2_{11}(c/c_1),$

$$\begin{aligned} c &= \exp(c_1 x + c_2 y), & c_1 &= G^2_{12} - G^1_{11} \neq 0, \\ c_2 &= G^2_{22}, & c_1 G^2_{12} &= c_2 G^2_{11}, \end{aligned} \tag{B.12}$$

with the geodesics:

$$\begin{cases} c_1\ddot{x} + (c_0c_2 - c_1^2)\dot{x}^2 = 0, \qquad c_0 = G^2_{11}, \\ c_1\ddot{y} + c_0\dot{x}(c_1\dot{x} + c_2\dot{y}) + c_2\dot{y}(c_0\dot{x} + c_1\dot{y}) = 0, \end{cases}$$

or

$$y'' + c_2(y')^2 + (c_1^2 + c_0c_2)y'/c_1 + c_0 = 0.$$

(2°) $L = |\dot{x}| \exp(A + \varepsilon\dot{y}/\dot{x}), \quad A = (G^1_{11} + \varepsilon G^2_{11})x + \varepsilon G^2_{12}y,$

$$G^2_{12} = G^1_{11} = c_1, \quad G^2_{22} = 0,$$

with the geodesics: $\ddot{x} + c_1\dot{x}^2 = 0, \quad \ddot{y} + c_2\dot{x}^2 + 2c_1\dot{x}\dot{y} = 0,$ *or*

$$y'' + c_1 y' + c_2 = 0, \quad c_2 = G^2_{11}.$$

B.5 $B^2(r,s)$ with Constant G^i_{jk}

Assume that G^i_{jk} are constant in (B.3). Then we have

$$rA + sB = rc_1x + sc_2y + c_0,$$

$$c_1 = G^1_{11}, \quad c_2 = G^2_{22}, \quad c_0 = \text{const.}$$

Hence, we have

$$|a^r b^s| = \exp(rc_1x + sc_2y + c_0).$$

Proposition B.4. *A Finsler space which belongs to $B^2(r,s)$ and has constant G^i_{jk} in a standard coordinate system is such that*

$$L = |\dot{x}^r \dot{y}^s| \exp(rc_1x + sc_2y + c_0),$$

$$c_1 = G^1_{11}, \quad c_2 = G^2_{22},$$

with the geodesics:

$$\ddot{x} + c_1\dot{x}^2 = 0, \quad \ddot{y} + c_2\dot{y}^2 = 0,$$

or

$$y'' = -c_2(y')^2 + c_1y'.$$

Bibliography

[Ant79] P.L. Antonelli, *The geometry of random drift. V. Axiomatic derivation of the WFK diffusion from a variational principle*, Adv. in Appl. Probab. **11** (1979), 502–509.

[Ant85] P.L. Antonelli (ed.), *Mathematical essays on growth and the emergence of form*, University of Alberta Press, Edmonton, Canada, 1985.

[Ant90] P.L. Antonelli, *Applied Volterra-Hamilton systems of Finsler type: Increased species diversity as a non-chemical defense for corals against the crown-of-thorns*, (R.H. Bradbury, ed.), Lecture Notes in Biomathematics, vol. 88, 1990, pp. 220–235.

[Ant91] P.L. Antonelli, *Finslerian Volterra-Hamilton systems in ecology*, Tensor (N.S.) **50** (1991), 22–31.

[Ant93] P.L. Antonelli, *On y-Berwald connections and Hutchinson's ecology of social interactions*, Tensor (N.S.) **52** (1993), 27–36.

[AB96] P.L. Antonelli and R. Bradbury, *Volterra-Hamilton models in the ecology and evolution of colonial organisms*, Series in Mathematical Biology and Medicine, Word Scientific Press, Singapore, 1996.

[ABKS93] P.L. Antonelli, R. Bradbury, V. Krivan, and H. Shimada, *A dynamical theory of heterochrony: Time-sequencing changes in ecology, development and evolution*, J. Biol. Systems **1** (1993), 451–487.

[ABL91] P.L. Antonelli, R. Bradbury, and X. Lin, *A higher-order predator-prey interaction with application to observed starfish waves and cycles*, Ecol. Modelling **58** (1991), 323–332.

[ABL92] P.L. Antonelli, R. Bradbury, and X. Lin, *On Hutchinson's competition equations and their homogenization: A higher-order principle of competitive exclusion*, Ecol. Modelling **60** (1992), 309–320.

[ACV80] P.L. Antonelli, J. Chapin, and B. Voorhees, *The geometry of random drift. VI. A random selection diffusion model*, Adv. in Appl. Probab., **12** (1980), 50–58.

[AE86] P.L. Antonelli and R.J. Elliott, *Nonlinear filtering theory for coral/starfish and plant/herbivore interactions*, Stochastic Anal. Appl. **4** (1986), 1–23.

[AES87] P.L. Antonelli, R.J. Elliott, and R.M. Seymour, *Nonlinear filtering and Riemannian scalar curvature* (R), Adv. in Appl. Math. **8** (1987), 237–253.

[AFK87] P.L. Antonelli, K.D. Fuller, and N. Kazarinoff, *A study of large amplitude periodic solutions in a model for starfish predation on coral*, IMA J. Math. Appl. Med. Biol. **4** (1987), 207–214.

[AIM93] P.L. Antonelli, R.S. Ingarden, and M. Matsumoto, *The theory of sprays and Finsler spaces with applications in physics and biology*, Kluwer Academic Publishers, Dordrecht-Boston-London, 1993.

[AK84] P.L. Antonelli and N. Kazarinoff, *Starfish predation of growing coral reef community*, J. Theor. Biol. **107** (1984), 667–684.

[AKR+88] P.L. Antonelli, N.D. Kazarinoff, R.E. Reichelt, R.H. Bradbury, and P.J. Moran, *A diffusion-reaction transport model for large-scale waves in crown-of-thorns starfish outbreaks on the Great Barrier Reef*, IMA J. Math. Appl. Med. Biol. **6** (1988), 81–89.

[AL90] P.L. Antonelli and X. Lin, *Bifurcation analysis of a coral-starfish model*, Mathl. and Comput. Modelling **13** (1990), 35–44.

[AM96] P.L. Antonelli and R. Miron (eds.), *Lagrange and Finsler geometry. Applications to physics and biology*, Kluwer Academic Press, Dordrecht, 1996.

[AS92] P.L. Antonelli and P. Sammarco, *Allelochemic interactions on Pandora Reef and The Great Barrier Reef: Stabilizing weak chaos in production*, Open Systems and Information Dynamics **1** (1992), 207–215.

[AS87a] P.L. Antonelli and R.M. Seymour, *Nonlinear filtering of Myxomatosis based on hormonal control of flea reproduction by the rabbit*, Stoch. Anal. and Appl. **5** (1987), 245–265.

[AS87b] P.L. Antonelli and R.M. Seymour, *Entropy production in stochastic Riemannian geometries with applications to chemical ecology*, Adv. in Appl. Math. **8** (1987), 254–280.

[AS88a] P.L. Antonelli and R.M. Seymour, *A model of Myxomatosis based on hormonal control of rabbit-flea reproduction*, IMA J. Math. Appl. Med. Biol. **5** (1988), 65–80.

[AS91] P.L. Antonelli and H. Shimada, *On 1-form Finsler connections with constant coefficients*, Tensor, N.S. **50** (1991), 263–275.

[AS88b] P.L. Antonelli and J.N. Skowronski, *Identification of states and parameters in a model of starfish predation on corals*, Math. Comput. Modelling **10** (1988), 17–25.

[AS77] P.L. Antonelli and C. Strobeck, *The geometry of random drift. I. Stochastic distance and diffusion*, Adv. in Appl Probab. **9** (1977), 238–249.

[AV83] P.L. Antonelli and B. Voorhees, *Nonlinear growth mechanics. I. Volterra-Hamilton systems*, Bull. Math. Biol. **45** (1983), 103–116.

[AZ94] P. L. Antonelli and T. J. Zastawniak, *Density-dependent host/parasite systems of Rothschild type and Finslerian diffusion*, Math. Comput. Modelling **20** (1994), no. 4-5, 117–129.

[AZ96] P.L. Antonelli and T.J. Zastawniak, *Curvature and production stability in Volterra-Hamilton systems of Finsler type*, Open Systems and Information Dynamics **3** (1996).

[AZ97] P.L. Antonelli and T.J. Zastawniak, *Stochastic Finsler geometry in the theory of evolution by symbiosis*, Dynamics of Continuous, Discrete and Impulsive Systems **3** (1997), no. 1, 1–18.

[Arn74] L. Arnold, *Stochastic differential equations: Theory and applications*, Interscience, New York-London, 1974.

[BL96] D. Bao and B. Lackey, *A Hodge decomposition theorem for Finsler spaces*, C.R. Acad. Sci., Paris, Ser. 1 **323** (1996), 51–56.

[Car34] M.E. Cartan, *Les espaces de Finsler*, Actualités, vol. 79, Hermann, Paris, 1934, 2nd edition, 1971.

[Car39] M.E. Cartan, *Les problèmes d'equivalence*, (Paris), Paris, 1939, pp. 113–136.

[Che43] S.S. Chern, *On the Euclidean connections in a Finsler space*, Proc. Nat. Acad. Sci. U.S.A. **29** (1943), 33–37.

[Che48] S.S. Chern, *Local equivalence and Euclidean connections in Finsler spaces*, Sci. Rep. Nat. Tsing Hua Univ. Ser. A **5** (1948), 95–121.

[Coo64] P.H. Cootner (ed.), *The random character of stock market prices*, MIT Press, Cambridge, Mass., 1964.

[Doo53] J.L. Doob, *Stochastic processes*, John Wiley and Sons, New York, 1953.

[Dou28] J. Douglas, *The general geometry of paths*, Ann. of Math. **29** (1927/28), 143–169.

[Dyn65] E.B. Dynkin, *Markov processes*, Springer-Verlag, Berlin, 1965.

[Eis27] L.P. Eisenhart, *Non-Riemannian geometry*, AMS Colloq. Publ., vol. 8, American Mathematical Society, 1927.

[Eis49] L.P. Eisenhart, *Riemannian geometry*, Princeton Univ. Press, Princeton, NJ, 1949.

[Ell82] R.J. Elliott, *Stochastic calculus and applications*, Springer-Verlag, New York, 1982.

[Elw82] K.D. Elworthy, *Stochastic differential equations on manifolds*, London Mathematical Society Lecture Notes Series, Cambridge University Press, Cambridge, 1982.

[Eme89] M. Emery, *Stochastic calculus in manifolds*, Springer-Verlag, Berlin-Heidelberg-New York, 1989.

[Fri75] A. Friedman, *Stochastic differential equations and applications*, Academic Press, New York-San Francisco-London, 1975.

[GS72] I.I. Gihman and A.V. Skorohod, *Stochastic differential equations*, Springer-Verlag, Berlin-Heidelberg-New York, 1972.

[GM83] F. Guerra and L.M. Morato, *Quantization of dynamical systems and stochastic control theory*, Phys. Rev. D **27** (1983), 1774–1786.

[Gra77a] R. Graham, *Covariant formulation of non-equilibrium statistical thermodynamics*, Z. Phys. B **26** (1977), 397–405.

[Gra77b] R. Graham, *Path integral formulation of general diffusion processes*, Z. Phys. B **26** (1977), 281–290.

[HAC+69] N. Hairston, J. Allan, R. Colwell, D. Futuyma, J. Howell, M. Lubin, J. Mathias, and J. Vandermeer, *The relationship between species diversity and stability: and experimental approach with protozoa and bacteria*, Ecology **49** (1969), 1091–1101.

[Has75] M. Hashiguchi, *On Wagner's generalized Berwald spaces*, J. Korean Math. Soc. **12** (1975), 51–61.

[Has86] M. Hashiguchi, *Some topics on Finsler geometry*, Conferenze del Seminario di Matematica dell'Università di Bari **210** (1986), 1–26.

[Hid80] T. Hida, *Brownian motion*, Springer-Verlag, Berlin-Heidelberg-New York, 1980.

[Hör69] L. Hörmander, *Hypoelliptic second order differential operators*, Acta Math. **119** (1969), 147–171.

[Hut47] G.E. Hutchinson, *A note on the theory of competition between social species*, Ecology **28** (1947), 319–321.

[Hut78] G.E. Hutchinson, *An introduction to population biology*, Yale University Press, New Haven, Conn., 1978.

[Huz49] V.S. Huzurbazzar, *On a property of distributions admitting sufficient statistics*, Biometrika (1949), 71–74.

[IH77] Y. Ichijyō.*Finsler manifolds modeled on a Minkowski space*, J. Math. Kyoto Univ. **16** (1976), 639–652.

[IW89] N. Ikeda and S. Watanabe, *Stochastic differential equations and diffusion processes*, second ed., North-Holland, Amsterdam-Oxford-New York, 1989.

[Itô87] K. Itô, *Differential equations determining Markov processes*, Selected Papers (Berlin), Springer-Verlag, Berlin, 1987.

[Itô44] K. Itô, *Stochastic integral*, Proc. Imp. Acad. Tokyo **20** (1944), 519–524.

[IM64] K. Itô and H.P. McKean, Jr., *Diffusion processes and their sample paths*, Academic Press, New York, 1964.

[KS88] I. Karatzas and S.E. Shreve, *Brownian motion and stochastic calculus*, Springer-Verlag, New York-Berlin-Heidelberg, 1988.

[Kol37] N. Kolmogorov, *Zur Umkehrbarkeit der statistischen Naturgesetze*, Math. Ann. **113** (1937), 766–772.

[Kop84] P.E. Kopp, *Martingales and stochastic integrals*, Cambridge University Press, Cambridge, 1984.

[Kry68] E. Kreyszig, *Introduction to differential geometry and Riemannian geometry*, Math. Expositions, **16**, Univ. Toronto Press, 1968.

[Kun83] H. Kunita, *Stochastic partial differential equations connected with nonlinear filtering*, in Nonlinear Filtering and Stochastic Control, Lecture Notes in Math. **972**, Springer-Verlag, New York/Berlin, 1983, 101–169.

[Kun90] H. Kunita, *Stochastic flows and stochastic differential equations*, Cambridge University Press, Cambridge, 1990.

[Kur73] T.G. Kurtz, *A limit theorem for perturbed operator semigroups with applications to random evolutions*, J. Functional Analysis **12** (1973), 55–67.

[LS77] R.S. Liptser and A.N. Shiryayev, *Statistics of random processes*, Springer-Verlag, New York, 1977.

[Mat66] M. Matsumoto, *A Finsler connection with many torsions*, Tensor, N.S. **17** (1966), 217–226.

[Mat82] M. Matsumoto, *On Wagner's generalized Berwald space of dimension 2*, Tensor, N.S. **36** (1982), 303–311.

[Mat86] M. Matsumoto, *Foundations of Finsler geometry and special Finsler spaces*, Kasheisha Press, Otsu, Japan, 1986.

[Mat94] M. Matsumoto, *Goedesics of two-dimensional Finsler spaces*, Math. Comput. Modelling **20** (1994), no. 4-5, 1–23.

[McK69] H.P. McKean, Jr., *Stochastic integrals*, Academic Press, New York-London, 1969.

[Med40] P.B. Medawar, *The growth, growth energy, and aging of chicken's heart*, Proc. Royal Soc. London **B129** (1940), 332–355.

[MR51] A.N. Milgram and P.C. Rosenbloom, *Harmonic forms and heat conduction in closed Riemannian manifolds*, Proc. N.A.S. **37** (1951), 180–184.

[MA94] R. Miron and M. Anastasiei, *The geometry of Lagrange spaces: Theory and applications*, Kluwer Academic Press, Dordrecht, 1994.

[Mol75] S.A. Molchanov, *Diffusion processes and Riemannian geometry*, Russian Math. Surveys **30:1** (1975), 1–63.

[Nag80] M. Nagasawa, *Segregation of a population in an environment*, J. Math. Biol. **9** (1980), 213–235.

[Nel67] E. Nelson, *Dynamical theories of Brownian motion*, Princeton University Press, Princeton, New Jersey, 1967.

[Nel85] E. Nelson, *Quantum fluctuations*, Princeton University Press, Princeton, New Jersey, 1985.

[Pin78] M. Pinsky, *Stochastic Riemannian geometry*, (New York) (A.T. Bharucha-Reid, ed.), Academic Press, New York, 1978, pp. 199–235.

[Pin92] M. Pinsky, *Inverse questions in stochastic differential geometry*, (Berlin-New York) (L. Chen, K. Choi, K. Hu., and J.H. Lou, eds.), Walter de Gruyter & Co., Berlin-New York, 1992, pp. 3–28.

[Rao60] B.R. Rao, *A formula for the curvature of the likelihood surface of a sample drawn from a distribution admitting sufficient statistics*, Biometrika **47** (1960), 203–208.

[Rie92] B. Riemann, *Über die Hypothesen, welsche der Geometrie zu Grunde liegen, Habilitationsvortrag*, Geom. Math. Werke (Leipzig), Leipzig, 1892, Reproduced by Dover Publications, 1953, pp. 272–288.

[RW90] L.C.G. Rogers and D. Williams, *Diffusions, Markov processes and martingales. Itô calculus*, John Wiley and Sons, Chichester-New York, 1990.

[Run59] H. Rund, *The differential geometry of Finsler spaces*, Springer-Verlag, Berlin-Göttingen-Heidelberg, 1959.

[Str71] D.W. Stroock, *On the growth of stochastic integrals*, Z. Wahr. verv. Gebiete **18** (1971), 340–344.

[Sza81] Z.I. Szabó, *Positive definite Berwald spaces. (Structure theorems on Berwald spaces)*, Tensor (N.S.) **35** (1881), 25–39.

[TW81] Y. Takahashi and S. Watanabe, *The probability functionals (Onsager-Machlup functions) of diffusion processes*, Lecture Notes in Math., vol. 851, Springer-Verlag, 1981.

[Var84] S.R.S. Varadhan, *Large deviations and applications*, SIAM Publications, Philadelphia, 1984.

[Wel80] R.O. Wells, *Differential analysis on complex manifolds*, Springer-Verlag, Berlin-New-York, 1980.

[Wic95] D. Wick, *The infamous boundary: Seven decades of controversy in quantum physics*, Birkhäuser, Boston, 1995.

[Wil71a] H. Wilbur, *Competition, predation and the structure of the Abystoma-Rana Sylvatica community*, Ecology **53** (1971), 3–21.

[Wil71b] H. Wilbur, *The ecological relationship of the salamander Abystoma Laterale to its all-female gynogenetic associate*, Evolution **25** (1971), 168–179.

[Wil79] D. Williams, *Diffusions, Markov processes and martingales. Foundations*, John Wiley and Sons, Chichester, 1979.

[YF80] H. Yasuda and M. Fukui, *On the curvature of the indicatrix bundle over a Finsler space*, Ann. Rep. Asahikawa Med. Coll. **2** (1980), 1–21.

[Yos49] K. Yoshida, *Brownian motion on the surface of the 3-sphere*, Ann. Math. Stat. **20** (1949), 292–290.

Index

68. M. Evans and J.-P. Vigier: *The Enigmatic Photon.* Volume 2: Non-Abelian Electrodynamics. 1995 ISBN 0-7923-3288-1
69. G. Esposito: *Complex General Relativity.* 1995 ISBN 0-7923-3340-3
70. J. Skilling and S. Sibisi (eds.): *Maximum Entropy and Bayesian Methods.* Proceedings of the Fourteenth International Workshop on Maximum Entropy and Bayesian Methods. 1996 ISBN 0-7923-3452-3
71. C. Garola and A. Rossi (eds.): *The Foundations of Quantum Mechanics – Historical Analysis and Open Questions.* 1995 ISBN 0-7923-3480-9
72. A. Peres: *Quantum Theory: Concepts and Methods.* 1995 (see for hardback edition, Vol. 57) ISBN Pb 0-7923-3632-1
73. M. Ferrero and A. van der Merwe (eds.): *Fundamental Problems in Quantum Physics.* 1995 ISBN 0-7923-3670-4
74. F.E. Schroeck, Jr.: *Quantum Mechanics on Phase Space.* 1996 ISBN 0-7923-3794-8
75. L. de la Peña and A.M. Cetto: *The Quantum Dice.* An Introduction to Stochastic Electrodynamics. 1996 ISBN 0-7923-3818-9
76. P.L. Antonelli and R. Miron (eds.): *Lagrange and Finsler Geometry.* Applications to Physics and Biology. 1996 ISBN 0-7923-3873-1
77. M.W. Evans, J.-P. Vigier, S. Roy and S. Jeffers: *The Enigmatic Photon.* Volume 3: Theory and Practice of the $\boldsymbol{B}^{(3)}$ Field. 1996 ISBN 0-7923-4044-2
78. W.G.V. Rosser: *Interpretation of Classical Electromagnetism.* 1996 ISBN 0-7923-4187-2
79. K.M. Hanson and R.N. Silver (eds.): *Maximum Entropy and Bayesian Methods.* 1996 ISBN 0-7923-4311-5
80. S. Jeffers, S. Roy, J.-P. Vigier and G. Hunter (eds.): *The Present Status of the Quantum Theory of Light.* Proceedings of a Symposium in Honour of Jean-Pierre Vigier. 1997 ISBN 0-7923-4337-9
81. M. Ferrero and A. van der Merwe (eds.): *New Developments on Fundamental Problems in Quantum Physics.* 1997 ISBN 0-7923-4374-3
82. R. Miron: *The Geometry of Higher-Order Lagrange Spaces.* Applications to Mechanics and Physics. 1997 ISBN 0-7923-4393-X
83. T. Hakioğlu and A.S. Shumovsky (eds.): *Quantum Optics and the Spectroscopy of Solids.* Concepts and Advances. 1997 ISBN 0-7923-4414-6
84. A. Sitenko and V. Tartakovskii: *Theory of Nucleus.* Nuclear Structure and Nuclear Interaction. 1997 ISBN 0-7923-4423-5
85. G. Esposito, A.Yu. Kamenshchik and G. Pollifrone: *Euclidean Quantum Gravity on Manifolds with Boundary.* 1997 ISBN 0-7923-4472-3
86. R.S. Ingarden, A. Kossakowski and M. Ohya: *Information Dynamics and Open Systems.* Classical and Quantum Approach. 1997 ISBN 0-7923-4473-1
87. K. Nakamura: *Quantum versus Chaos.* Questions Emerging from Mesoscopic Cosmos. 1997 ISBN 0-7923-4557-6
88. B.R. Iyer and C.V. Vishveshwara (eds.): *Geometry, Fields and Cosmology.* Techniques and Applications. 1997 ISBN 0-7923-4725-0
89. G.A. Martynov: *Classical Statistical Mechanics.* 1997 ISBN 0-7923-4774-9
90. M.W. Evans, J.-P. Vigier, S. Roy and G. Hunter (eds.): *The Enigmatic Photon.* Volume 4: New Directions. 1998 ISBN 0-7923-4826-5
91. M. Rédei: *Quantum Logic in Algebraic Approach.* 1998 ISBN 0-7923-4903-2
92. S. Roy: *Statistical Geometry and Applications to Microphysics and Cosmology.* 1998 ISBN 0-7923-4907-5

Fundamental Theories of Physics

93. B.C. Eu: *Nonequilibrium Statistical Mechanics.* Ensembled Method. 1998
    ISBN 0-7923-4980-6
94. V. Dietrich, K. Habetha and G. Jank (eds.): *Clifford Algebras and Their Application in Mathematical Physics.* Aachen 1996. 1998 ISBN 0-7923-5037-5
95. *Not yet known*
96. V.P. Frolov and I.D. Novikov: *Black Hole Physics.* Basic Concepts and New Developments. 1998 ISBN 0-7923-5145-2; PB 0-7923-5146
97. G. Hunter, S. Jeffers and J-P. Vigier (eds.): *Causality and Locality in Modern Physics.* 1998
    ISBN 0-7923-5227-0
98. G.J. Erickson, J.T. Rychert and C.R. Smith (eds.): *Maximum Entropy and Bayesian Methods.* 1998 ISBN 0-7923-5047-2
99. D. Hestenes: *New Foundations for Classical Mechanics (Second Edition).* 1999
    ISBN 0-7923-5302-1; PB ISBN 0-7923-5514-8
100. B.R. Iyer and B. Bhawal: *Black Holes, Gravitational Radiation and the Universe.* Essays in Honor of C. V. Vishveshwara. 1999 ISBN 0-7923-5308-0
101. P.L. Antonelli and T.J. Zastawniak: *Fundamentals of Finslerian Diffusion with Applications.* 1999 ISBN 0-7923-5511-3

KLUWER ACADEMIC PUBLISHERS – DORDRECHT / BOSTON / LONDON

GPSR Compliance

The European Union's (EU) General Product Safety Regulation (GPSR) is a set of rules that requires consumer products to be safe and our obligations to ensure this.

If you have any concerns about our products, you can contact us on ProductSafety@springernature.com

In case Publisher is established outside the EU, the EU authorized representative is:

Springer Nature Customer Service Center GmbH
Europaplatz 3
69115 Heidelberg, Germany

Batch number: 10755292

Printed by Printforce, the Netherlands